T0278184

The New Empire of AI

The New Empire of AI

The Future of Global Inequality

RACHEL ADAMS

polity

First published in 2025 by Polity Press

Polity Press
65 Bridge Street
Cambridge CB2 1UR, UK

Polity Press
111 River Street
Hoboken, NJ 07030, USA

ISBN-13: 978-1-5095-5309-9

A catalogue record for this book is available from the British Library.

Library of Congress Control Number: 2024937085

Typeset in 11 on 14pt Warnock Pro
by Cheshire Typesetting Ltd, Cuddington, Cheshire
Printed and bound in Great Britain by CPI Group (UK) Ltd, Croydon

For further information on Polity, visit our website:
politybooks.com

For Yazeed

Contents

Acknowledgements   ix

Prologue   xi

Introduction: The AI Divide   1
    Empire's Inequalities   11
    Empire, Old and New   15
    The Majority World   18

Chapter 1: A New World Order   23
    The Dawn of AI   27
    Rulers of the World   34
    The Racial State   46

Chapter 2: The Cost of AI   54
    Billion-Dollar Enterprises   58
    Colonial Economics   69
    Technology Diffusion   75

Chapter 3: The Material World of AI   79
    AI's Beginnings   83
    The Battle for the Heart of Darkness   89
    The Electric Storm   94

Chapter 4: The New Division of Labour   100
   Equal to or Somewhat Better Than an Unskilled Human   102
   Rationalizing Informality   111
   I Am Not a Robot   116
   Workers, Resist!   120

Chapter 5: Fit for What Purpose?   123
   Blood Money   127
   Biometric Empires   130

Chapter 6: One Language to Rule Them All   139
   A New Frontier   140
   Dominant Worldviews   145
   In the Margins   149

Chapter 7: The Way Out   154
   The Limits and New Horizons of AI Ethics   158
   What Governments and Governance Should Do   166
   A Different Kind of Leader   175

Coda: The New Politics of Revolution   178
   This Will Affect Us All   179
   The Time for Action is Now   180

Notes   183
Key Readings   196
Index   202

Acknowledgements

This book began as an article written to bring together two fields of thought in which I have been deeply invested. The first field is that of postcolonialism, decoloniality and critical race studies. I have been lucky enough to work with an incredible group of scholars – largely here in South Africa – whose work has profoundly influenced my own. My thanks here go to Joel Modiri, Sanele Sibande, Tshepo Madlingozi, Sabelo Ndlovu-Gatsheni, Pramesh Lalu, Jaco Barnard-Naudé and, in particular, Crain Soudien, whose friendship and encouragement have meant a great deal to me. The second is the emerging field of AI ethics. I'd like to thank especially to Stephen Cave, Nora Ni Loideain and Kanta Dihal, who have been instrumental to my work in this space.

As I sought to bring these two worlds together and to write this book, I have benefitted greatly from the kindness and support of, and long conversations with, many friends and colleagues to whom I would like to extend my deepest thanks: Matthew Smith, Paul Plantinga, Fola Adeleke, Mark Gaffley, Nico Grossman, Kelly Stone, Melanie George, Naila Govan-Vassen, Rosalind Parkes-Ratanshi, Alan Blackwell, Michael Gastrow, Temba Masilela, Urvashi Aneja, Jantina de Vries,

Kiito Shilongo, Andrew Merluzzi, Shachee Doshi, Zameer Brey, Divine Fuh and Jane Taylor.

Much of the work I have led and been engaged in over the past few years has allowed me to explore more deeply the implications of AI on inequality. This work would not have been possible without the generosity of the International Development Research Centre (IDRC) of Canada, which sponsors both the African Observatory on Responsible AI and the Global Index on Responsible AI. I'm grateful for the collegiality and conviviality that has characterized all my engagements and partnerships with the IDRC.

I remain deeply grateful to my editor at Polity, Jonathan Skerrett; I could not have asked for a more diligent and understanding partner in getting this book over the line. I am indebted to the incredible team at Polity for all their support.

I also thank Crain Soudien, Stephen Cave, Aubra Anthony, Aisha Sobey and Shakir Mohamed for their critical input and comments on various chapters and drafts of this book, all of which greatly helped to make the story I sought to tell more compelling.

My families on both sides of the world have provided crucial support as this book was written. A big thank you to you all, and especially to my dear children, who now have an outsize interest in AI. I am especially grateful to my parents, whose close readings of the text of this book were invaluable.

Lastly, thank you to my husband, my partner in life and thought, who has been with me every step of the way.

Prologue

When I was first ruminating on the idea of writing a book about the relationship between artificial intelligence (AI) and the history and contemporary conditions in Africa and across the so-called global South, I sought to describe an issue that I could see was bubbling beneath the surface. Having spent the earlier years of my career working on human rights in South Africa, I became aware that the capacity of African states to fulfil their duties in protecting the rights of their citizens was being undermined by the rise of new digital technologies and by the forms of global power in which they were entangled. In 2018, having lived in South Africa for ten years, I moved with my family back to the United Kingdom, to complete a postdoctoral training at the University of London. At the time AI was fast becoming a buzzword. The House of Lords was set to publish its first major report on AI, entitled *AI in the UK: Ready, Willing and Able?*, while China's AI capabilities were quickly ascending, prompting fears around the rise of a new global power not of the West's own making.

While some anxiety existed around AI's effects on democratic stability, as the Snowden revelations and the Cambridge Analytica scandal had just recently demonstrated the impact

and reach of new technologies on individual rights and free-
doms, these concerns were not linked to their broader global
resonance. Where issues of bias in AI were acknowledged, they
were considered system-level concerns, to be solved through
better programming. They were not connected to the structural and historical conditions out of which inequality and
discrimination have arisen in the world.

In all of this, the position and the fate of the global South
and of the African continent in particular were simply being
ignored. These places mattered little to the exciting new world
into which AI was ushering us. During my postdoc in 2018, I
became increasingly concerned about a new global agenda that
failed to include the majority of the world but clearly seemed
dependent on it, whether for the extraction of resources or for
the creation of new markets.

In 2019 we moved back to South Africa, where my work
focused on understanding the impacts and implications of AI
outside the West. One of the major areas of my work was a
platform that I established in order to promote African experiences and expertise in the global discussions and debates about
AI – which, in any event, were affecting the continent. This
platform, the African Observatory on Responsible AI, was supported and funded by the International Development Research
Centre of Canada and took on a life of its own. As the platform
grew, our work mushroomed from research on the use and
impacts of AI in Africa to advising and training African policymakers, working directly with innovators across the continent
who built humane and beneficial AI technologies, and engaging with regional bodies and multilateral institutions. There
was much work to be done – and still is.

There is also something important about the African and
South African perspective that needs to be heard. South Africa
is consistently rated the most unequal country in the world.
This has everything to do with its complex colonial history, as
the state-sanctioned racial segregation of apartheid imposed

the worst form of racial inequality. After 1994, in the years that followed the demise of apartheid and the establishment of democratic majority rule, South Africa has continued to champion the cause of creating an equal society and world, in a global order where its sovereignty to defend the best interests of its citizens is sometimes impeded. Here, too, there is much work to be done to address the inequality and related social ills that continue to pervade South African society. But the tenacity to do that work exists in the spirit of the people and communities who keep fighting for justice.

South Africa's experiences in trying to build an equal and prosperous society are important lessons for the world. And its efforts are not just internal to the country. In late 2023, South Africa led the world in seeking justice for the horrors that Israel was committing against the Palestinian population in Gaza. Indeed, recent reports have brought to light Israel's appalling use of automated AI-driven technology in the war on Gaza with programs with cruelly satirical names such as 'Lavender' and 'the Gospel'.[1] But South Africa's protest against Israel was also a cry to all nations to uphold the sanctity of the international system of human rights and humanitarian law in a frightening and fragmenting world. As AI's planetary power expands, we will need these global systems more than ever.

As the writing of this book continued, the issues I set out to convey have intensified into urgent crises that require urgent action. The trajectory of worsening global inequality that we are confronting as a global society and in which AI is fully implicated is not just a trend. At the centre of it are human lives and livelihoods, ambitions and dreams. People, particularly those who live in the majority world, are paying the price for this new empire of AI, as I call it. The introduction to this book and the opening of Chapter 4 offer a number of vignettes of human stories related to AI. While these stories are presented as hypotheticals, in non-identifying form, they are real accounts I've heard time and time again.

This book is not meant to cause despair. It is intended to call us all to action – to hold to account this new global power that reigns among us. I hope this book inspires a new commitment to our collective global humanity and to what, together, we are capable of doing.

Introduction

The AI Divide

In a shack outside the city of Johannesburg, a young woman considers her options. She lives with her son in an iron shack hastily built out of discarded material from building sites across Johannesburg's leafy suburbs. She has 20 rand (R20) left – about $1 – after collecting her monthly childcare grant, which is normally around $115 per month. She had recently signed up for a new loan agreement in order to pay for childcare for her son, so that she may be able to undertake occasional work as a domestic cleaner. This work pays her about R180 ($9) a day. But the loan's repayment terms were punishing, and she has been unable to pay back in time what she owed. Unbeknownst to her, the parent company of the agency she borrowed money from is the same company that distributes her monthly grant. As this company has access to her data, the loan repayments are automatically deducted from the grant before she even sees the money.

* * *

Upon arrival at Dulles International Airport, Washington, a man is approached by security personnel. He is returning home from a trip to Pakistan. He's been visiting his family. His facial features and Muslim name prompt an AI-automated alert within the airport's security systems, causing the ground authorities to detain him for interrogation in an airless room within the airport compound.

* * *

A young man of Tigrayan descent is on the run. A social media post has gone viral. It has been authored by an Amhara general – a leader of the regional armed forces that are conducting a campaign of ethnic cleansing against Tigrayans. The post suggests that a group of Tigrayans live in a nearby village. It is written in local Amharic and has not been deleted by Facebook's AI-powered hate-speech detector.

* * *

The screen lights up with a new notification. A teenage girl in Malaysia is manning the computer until her father returns. She knows that she has just 15 minutes to respond to the request before the task is offered to someone else. She clicks on the link, which opens up a series of images displaying visceral scenes of indescribable violence. Quickly, she labels the images 'gratuitous violence' and closes the task. Her work will stop these images from being used in the training data of advanced image-generation AI, keeping this technology 'clean' for its end-users.

* * *

Over the past few years, a sophisticated set of technologies have been created and put to use around the world. They are called 'artificial intelligence' and, as their name suggests, these technologies seek to mimic the capabilities of human intelligence, particularly learning, logic, decision-making, and the recognition of speech, objects, and images. In its material existence within the world – a materiality that is largely hidden from the public eye – AI encompasses a vast and expanding network of computer-based technologies that require an interminable input of data and seek to display human-like intelligence and functioning. It involves a mega industry that operates at a planetary scale, with a footprint in every corner of the globe and an infrastructure that is both subterranean and celestial, extending to the deepest parts of the earth's oceans and circulating into the outer atmosphere of our cosmos. And it enfolds some of the most ambitious pursuits of the human species: not just to create life, but to create the meta-intelligence from which new planetary species and forms can be moulded and brought into being.

Around the world, AI is impacting the lives of ordinary people. For some, things are getting easier. They outsource mundane daily tasks to AI-powered assistants and use new generative AI capabilities that enhance work performance and productivity in their white-collar professions. AI-powered tools equip their children with personalized educational tools, tailored to the learning needs and the pace of each child. And when they are unwell, their doctors draw on AI-driven precision medicine to provide personalized diagnoses and prescriptions. One day, these people will have the opportunity to consider their digital afterlife and how advancements in AI could fuel the continuation of their character and voice long after their death.

But, as AI serves to accelerate prosperity and wellbeing in those places where it is produced and readily integrated into social and economic life, other people elsewhere pay the price.

AI makes life harder for many people across the world – just as it does for the people in the chapter-opening vignettes. For example, AI can hinder people's ability to meet their most basic needs. Access to critical public services such as social assistance programs is increasingly mediated through AI technologies. To date, many of these programs have failed to meaningfully lift people out of poverty; they entrench instead poverty traps that become harder and harder to escape. The new forms of work created by the AI industry and value chain are also failing to provide avenues for human flourishing. They are often exploitative, dependent on the economic vulnerability of people across the world who have little choice but to undertake precarious tasks for which only very basic wages are paid. As the use of meAI increases around the world, so does the risk that these systems will produce unjust outcomes for people who are most vulnerable to gendered, racial, or ableist bias or most in need of the benefits these technologies might bring.

To date, conversations around the risks and limitations of AI and its future have largely centred on western experiences and evidence. While the question of bias is acknowledged, it is treated as a system-level issue, not as a global structural reality. Where fears of job losses and displacements arise, they are unconnected to the precarious digital labour that fuels the AI industry or to the new global divisions of labour that are taking shape. And where experts weigh in on the long-term risks of AI, the risk of rising levels of global inequality barely features.

Yet the countries that are left out of these conversations play an integral role in the production of this industry: a role that is largely unseen, and certainly not fairly valued. The people who live in these places put in their labour to build and maintain the vast lakes of data upon which AI draws for its resources. Their lands provide the raw earth materials that build the hardware upon which and through which AI operates. Their societies provide the testing ground for trialing new technologies in spaces considered less important, where the collateral damage goes unseen.

In societies across the majority world where AI does exist, it is often ill suited to local needs – or else it is rolled out without the proper safeguards and institutional oversight that these young democracies so sorely need in order to ensure that the human rights of their citizens are not harmed or put at risk. Southern governments turn increasingly to AI solutions as quick fixes to intractable developmental problems such as the gap between supply and demand, and they do so for access to government services. But very often these systems function as poverty traps, failing to lift people out of poverty. In the slow violence of poverty, the marginalized and oppressed live in a constant state of war, fighting against systems that are increasingly harder to see and understand. Crucially, too, where AI is used to distribute resources, it can create divisions between communities, as can automated propaganda machines and sophisticated profiling techniques that function at once to polarize and to depoliticize the targeted people and communities.

The global auditing firm PricewaterhouseCoopers (PwC) attempted to put a price on the value AI would bring to national economies and to GDP growth. In a seminal report published in 2017, just as AI was seeping into global focus, PwC released 'Sizing the Prize', a piece that boasted that by 2030 AI would contribute $15.7 trillion to the global economy (see Figure 1). China, North America, and Europe stand to gain 85 per cent of this prize. The remainder is scattered across the rest of the world, with 3 per cent predicted for Latin America, 6 per cent for the region the report terms 'developed Asia', and 8 per cent for the entire bloc of 'Africa, Oceania and other Asian markets'. On the global map upon which the AI prize is split, Africa's potential gains are simply not there. Instead, this large continental mass, home to 18 per cent of the world's population, is lumped with Oceania and the other less developed regions of Asia, their collective icon of growth sitting above Australia.

Other global agencies have similarly developed statistical models for the future of a global society. Through research examining the impact of Covid-19 on extreme poverty (see Figure 2), the World Bank has forecast that by 2030 – during

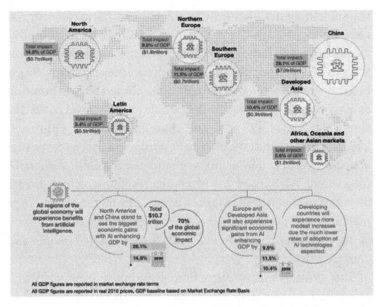

Figure 1   PricewaterhouseCoopers' 'Sizing the Prize' figure and
the invisible African continent (estimated GDP gains of AI to the
global economy)
Source: PwC, 2017.[1]

the same period in which the global economy will see gains
of $15.7 trillion from AI – 90 per cent of the world's poor will
live in sub-Saharan Africa. In fact, sub-Saharan Africa is set
to be the only region of the world where extreme poverty will
increase, as the rest of the world is set to experience significant
drops in the number of those who live well below the poverty
line. Added to this, with Africa's growing youth population
and the dwindling population numbers in European countries,
it is estimated that by 2050 a quarter of the world's population
will be African.

Little attention has been paid either to the potential for
AI to critically worsen the state of global inequality or to the
linkages between the enormous economic growth that AI will
deliver to particularly privileged zones and the worsening

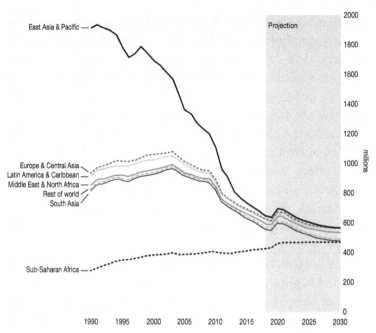

Figure 2   Number of the extremely poor (in millions), by region,
1990–2030
Source: Yonzan, Lakner, & Mahler, 2020.[2]

of extreme poverty in postcolonial contexts, especially sub-Saharan Africa. Nor is this division static. One of the defining features of AI is that its rate of development and adoption is exponential. AI will simply teach itself to be better and more efficient; it is optimized to continuously self-improve. Critically, this means that, if AI is not redirected towards addressing global inequality and is left to bring wealth and prosperity to high-income countries increasingly, the global inequality gap is simply going to widen at an exponential rate into a more and more unequal future.

It is increasingly recognized that the benefits of AI are not evenly distributed; the policy response to this state of affairs is to affirm that efforts are needed to support those 'left behind' and help them 'catch up'. This is reflected too in the AI for

Good movements, which assume that the problem is not AI, but simply how it is used. AI used for the right reasons, for good, will fix everything. This kind of narrative is an easy extension of the idolization of AI that has arisen in recent years: a position that assumes that AI will produce positive net benefits for humanity even if it has not yet, and that it represents the pinnacle of enlightened scientific discovery and applied human reason. This is a new version of the old trickle-down effect. Within such framing, the only goal is to advance AI further and distribute it more widely. There is very little room to ask: do we want AI to play such a dominant role in our societies? And is AI really benefiting – or going to benefit – all of us?

The present book is concerned with creating the space to ask, and begin to answer, these questions. It is written on the back of over 15 years of living and working in South Africa and across the African continent, where the problem and lived reality of inequality is ever present. As I will describe in this book, the risks of AI are higher in places outside the West. This is so for several reasons – from widespread job displacement and precarious and temporary work 'gigs' to human rights atrocities in the AI supply chain and to biased or useless AI systems that further marginalize nonwestern groups. What's more, in many of the places across the global South, the institutional mechanisms that might ordinarily protect individual rights and citizens' interests are either failing or unavailable, as they deal at full capacity with more fundamental social issues.

In 2017 a story broke out in South Africa that fundamentally impacted public perceptions of the use of technology on a mass scale in the public sector, and consumed many of those who work in these areas, myself included, for a number of years.

The South African Social Security Agency is responsible for providing cash benefits to just under 50 per cent of South

Africans, the majority of whom are entirely dependent on this grant for their livelihoods. In 2012 this agency had entered into a contract with Cash Paymaster Services to distribute social grants to around 17 million beneficiaries across the country. The contract ran from 2012 to 2017, during which time the parent company of Cash Paymaster Services, Net1, established a number of subsidiary businesses. Net1 gave these companies access to the banking details, bank accounts, and grant beneficiary information of all the grant beneficiaries Cash Paymaster Services was servicing. The companies in question used this information to profile potential clients and onsell predatory financial services. Loans at extortionate interest rates were sold to South Africa's most vulnerable people, who were surviving off grants of less than $100 a month. Beneficiaries were subject to various kinds of automated decision-making (a kind of precursor to the more complex AI systems we find today) to assess the terms of credit on offer. And, because of the collusion between Cash Paymaster Services and the other subsidiaries of Net1, automated deductions were made from the accounts of grant beneficiaries as soon as their grant payment was released.

Reporting from GroundUp, a South African investigative journalism group, described how a mother came to collect her child benefit grant – which at the time was R350 (less than $20) – but her balance after the loan deductions was a mere 26 cents, an amount that the cash point could not even dispense. Many others faced similar hardships as a result of the predatory practices of Net1 and its subsidiaries.

Today these kinds of systems are driven by AI. In fact the upgrade to the Cash Paymaster Services system was a program called GovChat, inconspicuously connected to Net1; and this program now integrates AI capabilities to provide advanced analytics. Such systems are rife with major imbalances of power that are hard to define and detect, and even harder to hold to account. In South Africa, while a high court judgment was handed down against Cash Paymaster Services, the company has gone into liquidation and no actual relief appears to have been provided to the many South Africans whose lives were so gravely affected by it.

But the public indignation was important. Part of what makes it so hard to tell the story of how AI perversely affects citizens in the majority world is that there is not enough public indignation against AI when issues occur. For not sufficient numbers of these stories of harm are coming to light, and without these stories in the public domain it becomes hard to detect where AI is not benefitting people and communities and to question the authority of AI and its supposedly benevolent mission in the world. When stories of the negative impact of AI across the globe do appear, it is because of a whistleblower – as in the case of Daniel Motuang, who blew the whistle on the treatment of Facebook and other Big Tech companies' content moderators in Kenya (an issue we will delve into more in Chapter 4) – or because of fastidious investigative journalism, such as that published by Rest of World, which circulates stories about technology's impact beyond the West. This paucity is also the reason why the stories that have arisen about the bias exhibited in the outcomes of AI systems hold such important lessons for understanding how AI is likely to impact societies across the majority world – and particularly stories of racial and gendered bias that are coming to light in droves across the western world. These stories point to a deeper underbelly of AI, an underbelly we will traverse in this book. For, like those in the West who experience discrimination in the face of AI, much of the majority world is effectively considered an anomaly within the logic of the AI system, which has been built on the western experience of the world. Nonwestern people, experiences, and language are barely represented in the datasets on which AI technologies are trained and from which AI systems interpret the world they encounter. From the perspective of this AI system, the western world as reflected in its training data is the only world: everything either conforms to it or is rejected.

From fragments of stories about AI's use and effects beyond the West, in combination with an analysis of key statistical trends and historical accounts and with the support of notes from my own experiences, accumulated over the years, of examining and working to address the effects of digital technologies and AI on African and majority world societies, a stark picture emerges of a deeply divided world. Just as AI serves to accelerate prosperity and wellbeing in those places where it is produced and readily incorporated into social and economic life, it does not manifestly improve lives anywhere else or for anyone else. Instead, as this book will argue, it deepens poverty, fractures community and social cohesion, and exacerbates divides between people and between groups. On this tangled tapestry, it becomes clear that, in the uneven distribution of AI's benefits, those who benefit do so *because* others are being used and harmed to produce AI and to sustain its relevance and reliability. And those who are being exploited and oppressed in the production and use of AI are the very same people who have historically been exploited and oppressed by global powers: women, people of colour, and citizens of the majority world.

Empire's Inequalities

All over the world we are facing rising levels of global inequality; these levels are the same as they were at the height of European colonialism, at the turn of the twentieth century. The World Inequality Report of 2022 gives us this picture:

> Global inequalities seem to be about as great today as they were at the peak of Western imperialism in the early 20th century. Indeed, the share of income presently captured by the poorest

half of the world's people is about half what it was in 1820, before the great divergence between Western countries and their colonies. In other words, there is still a long way to go to undo the global economic inequalities inherited from the very unequal organization of world production between the mid-19th and mid-20th centuries.[3]

Inequality takes many forms. It is not a monolithic phenomenon and produces radically different experiences in different groups and individuals. Global statistics allow us to gain a bird's-eye view of how rampant different factors of inequality are across diverse groups and regions of the world. They will, however, be only a proxy for the real life experiences that any one individual may live through and withstand. Inequality conditions any one individual's ability to live a life of their own choosing and to reach their full potential. But, while the *experience* of inequality lies at the level of the individual, social inequality and economic inequality are structural phenomena. They are produced by forms of power that exist at given points in time whereby decisions are made that favour one group while oppressing or marginalizing another.

Evidence on AI systems that have displayed critical racial and gendered biases abounds. Tendayi Achiume, who was appointed by the United Nations as a Special Rapporteur to understand contemporary forms of racism and racial discrimination, declared in her report to the UN Human Rights Council that emerging digital technologies such as AI were sharpening inequalities along racial, ethnic, and national origin lines.[4] Stories like the ones included among this chapter's vignettes – about people wrongfully detained or denied access to financial services or benefits because of AI systems whose codes incorporate the racial biases of the world around us – are emerging around the world. A study published by the US National Institute of Standards and Technology, the key government body for measuring the performance of technologies against industrywide standards, demonstrated that, out of 189 AI-driven facial recognition technologies it reviewed, non-white faces were misidentified between 10 and 100 times more than white faces.

An important body of work, led largely by women of colour, examines the relationship between AI, digital technologies, and the production of new forms of racism and exclusion. Ruha Benjamin examines the uses of race in the history of technology, deftly exploring how race is itself used as a technology for oppressive ends, to support white supremacy.[5] The work of Safiya Umoja Noble, a co-founder of the Center for Critical Internet Inquiry at the University of California, exposed the deep levels of stereotyped discrimination that are at work in Google's search algorithms.[6] Writing specifically on the racialized histories of surveillance technologies, Simone Browne argues that these technologies are enacted on the bodies of people of colour, functioning to reproduce and reinscribe racialized hierarchies, categorizations, and social conditions.[7] This is how the precursors of AI were trying to manage and contain black bodies, and they assumed intention and conduct from a mere 'reading' of the appearances of the body.

AI has demonstrated an equally appalling performance on gender equality. AI-driven recruitment systems have been found to downgrade the ranking of CVs that contain references to women's colleges or women's rights advocacy. AI-powered assessments for credits and loans have generated different outcomes for men and women with the same financial profiles. At the same time, online advertising has used algorithmic profiling to show women and people of colour lower-paid and less prestigious job adverts. In fact the AI industry is dominated by men, only around 12 per cent of leadership positions being occupied by women. A recent survey revealed that a staggering 73 per cent of the women who work in tech have experienced gender-based bias that ranges from favouritism towards male colleagues to sexual harassment.[8]

These concerns have largely been raised in relation to evidence that has come to light from western contexts. Applied to the majority world, the racial and gender biases that AI has exhibited become even more acute. Facial recognition AI systems applied in countries where the majority of the population is non-white are risky, to say the least. And in many contexts across the majority world, women are even more marginalized and disempowered. Will AI help ordinary people to live better lives, or will it make their lives worse?

European colonialism produced the most profound forms of inequality between people and places that continue to structure our contemporary condition. Race and gender, the most pervasive categories of inequality, are both linked critically to European colonialism. Race was colonialism's central creed: an invented marker of human difference and worth. And gender – particularly western binary notions of gender and traditional gender norms – was fixed by colonialism in huge areas across the world.[9]

To date, there has been no systematic treatment of the relationship between colonial histories and the AI divide, while the relationship between AI and global inequality goes almost completely unacknowledged.[10] I hope that this book can contribute to a broader understanding of the relationship between AI and new and old forms of global inequality. Fundamentally, my question is this: why does AI seem to sharpen inequalities globally, while its bigwigs profess that is a technology for humanity? Do the new patterns of inequality that AI has brought about, or made manifest, bear any connection with the historical production of inequality engendered by European colonialism? And are they connected with the rising levels of inequality we are seeing the world over?

In order to answer these questions, I engage with perspectives and accounts of AI from the majority world and examine some of the key trajectories and trends of the industry from within a historical frame. This frame allows us to understand the social, political, and economic conditions that gave rise to an unequal world; it is a history that is profoundly connected to that of European colonialism and to the seizure of land and resources from places across Africa, Asia, and the Americas. Modern capitalism began in earnest in 1492, with the arrival of Christopher Columbus on the coast of the new world of the Americas, and even earlier, in 1444, as Henry of Portugal shipped slaves over from Guinea.[11] In understanding how global inequality arose within the history of modernity and capitalism, we can begin to find an expression for the new ways in which old forms of power continue their work of AI de-equalization – and to identify their structural relationship.[12]

Empire, Old and New

Technology has long been used for imperial advantage and as a tool by which colonizers exert power over others. The compass, for instance, played a crucial role in the establishment of European empires. In fact the history of this technology dates back to the Mongolian Empire, which during the thirteenth and the fourteenth centuries united the enormous landmass of Asia into the largest contiguous territory in imperial history. This vast political beast established the major trade routes that connected the East to the West; and, in doing so, it played a crucial role in facilitating the progress of globalization and early capitalism. The Mongols then brought to Europe technologies such as the compass, which had been invented in ancient China and became an essential tool in the rise of the next great empire: that of Europe, founded through passage across the seas. Over the course of the fourteenth and fifteenth centuries, the nations of Europe regularized transatlantic travel through the rise of military and maritime capabilities, linking

the continent of Africa with that of the Americas and insti-
tuting a system of global trade in human slaves. The empires
of Europe arose on the back of the transatlantic slave trade,
formalizing proper colonialism as the takeover of foreign land
by sovereign force. Whether it be the might of the Europeans'
firearms against indigenous African weaponry or the more
advanced warships used in the British takeover of Singapore,
technology has extended the reach and capabilities of human
power, often with devastating effect.

But the new does not follow exactly in the footsteps of the
old. The new forms of power, bound up with AI today, do
not neatly continue and parallel historical European colonialism. The lords and geographies of imperial power have shifted.
Europe is no longer the centre of global power, and Big Tech has
come to occupy a position of planetary supremacy that rivals
and surpasses that of most nation states worldwide. What's
more, a new global power is arising in the East, prompting new
geopolitical anxieties and strategies. And AI has everything to
do with it. How, then, do the histories of coloniality help us to
understand and reckon with this new world order?

As we explore the global relations of power in which AI
operates and to which it contributes, together with their
human impact – their impact on the lives of ordinary people
such as those in the vignettes at the beginning of this chapter –
I will show how global society today is confronting a far greater
modality of power than in the time of the great empires of
Europe. I will call this modality the new empire of AI.

Together, the great houses of AI – Microsoft, Apple,
Amazon, Alphabet (the parent company of Google), NVIDIA,
and Meta – own almost 20 per cent of the global economy.
Everywhere AI's stocks are up; its rate of growth is unprecedented. In fact, what makes AI so remarkable and so difficult
to parse in relation to previous technologies is that it can
self-improve (through a technique called 'machine learning') and beget the set of technologies that will come after it.

Technological progress is now no longer an entirely human feat.

But its rate of growth also means that AI is quick to leave others behind. As the AI industries boom in the West and in the Far East, the gap between countries grows wider and wider, at a speed that is only accelerating.

The new empire of AI is driven by the logic of expansion and of exponential, perpetual growth. It depends on practices of extraction from the very same places that it seeks to expand into and conquer. It is abstract and invisible: clouds in the ether that are hard to see, and even harder to resist. Its effects are a deepening of the divides of our world, a huge fissure burgeoning between those who benefit and those who pay. And the divides are multiple. While this book is primarily concerned with the divides between countries that AI is worsening, those that AI amplifies within countries and between groups and communities are equally concerning. But this is nothing new to imperial power, for which 'divide and conquer' has repeatedly been an efficient way of preventing revolutions.

There is another feature of this new empire that makes it bear resemblance to the empires of old. In Britain, in the time of the British Empire, especially during the long reign of Queen Victoria, there was a certain atmosphere and sensation that surrounded the idea of empire. In the public imagination, the British Empire was a thing of glory and an opportunity like no other. Here was the chance to stake one's claim to lands rich in gold and rubber and to embark on surefooted missions of charity, proselytizing for the wonders of European modernity in those new economic districts. The empire offered a life of luxury to the colonialist, in a way that is not dissimilar to the promise of AI today.

And, just as in the old empires of Europe, the new empire of AI validates and justifies its mission as one of bettering all humanity. More than anything, AI is being couched as a global opportunity, a promise at the feet of humankind itself, to extend into a new world space of ease, efficiency, and fulfilment made possible by a powerful yet submissive digital intelligence. The CEO of Google DeepMind – a leading AI firm that develops AI technologies that will drastically speed up drug discovery – has stated that AI will deliver us from our most intractable challenges. It is, he says, a meta-solution to any problem, including global poverty. In October 2023 Elon Musk declared, in conversation with Rishi Sunak, then the UK prime minister, that AI is ushering in an age of abundance and soon we will no longer be worried about universal basic income, as everyone will enjoy a universal high income. These promises reflect the naive allure of the British Empire in its heyday. But at the same time they serve to distract from the very real hardships that AI is causing and from the deep structural risk of sleepwalking into a world of even greater inequality.

The Majority World

In this book, my choice of the term 'majority world' is deliberate. Over the past few decades, various terms have been used to describe the regions of the world that pale in comparison to the privilege and wealth of the West: 'third world', then 'underdeveloped' and 'developing' countries, then again 'industrializing world'. By the year 2000, the Millennium Development Goals had been established to cement global consensus around the task of international development. In particular, these goals helped countries and international development organizations reduce rates of infant and maternal mortality, as well as preventable deaths from HIV/AIDs and tuberculosis. In 2015, the Sustainable Development Goals were launched, which reaffirmed a global commitment to development across 17 goals and 169 target areas.

But in the twenty-first century the overemphasis on development has so far failed to capture the political status and significance of developing nations and, in particular, the role these countries play in the globalization machine. It was at that point that the idea of the global South became popular. While broadly used as a geographic designation for countries south of the Equator, it was also a means of referring to the role of southern nations in the global supply chains. These countries provided the cheap labour, land, and resources that drove the engine of globalization. With this, the interrelationship between the West and the rest was partially recognized, even without the full acknowledgement of the tragedy that European colonialism caused to the colonized lands and people. The overarching narrative was the old development discourse, which instead placed the emphasis on the benevolent role of the North in helping the South to progress towards proper (western) civilization.[13]

More recently, the notion of a majority world has become more prominent. The term bears recognition of the fact that the people thus described, despite being politically and economically at the margins of global society, represent the vast majority of the world's population – in fact, about 85 per cent. Thus framed, the concept throws into question the morality of an elitist global order that serves less than 15 per cent of the human race. For Shahidul Alam, the Bangladeshi author, photographer, and social activist, 'majority world' defines this global community in terms of what is in it, rather than what it lacks. It is also a political statement: the West is outnumbered.[14]

For those of us who have witnessed South Africa's emergence from the racial totalitarianism of apartheid to a new democratic dispensation, the politics of the majority has been a crucial concept. Very often, the oppressed and vulnerable people of the world are considered to be the minority. We speak of minority groups and identities, and of minorities politics. Consideration for these people is treated as a paternalistic add-on, superfluous to the responsibilities of democracy or the free functioning of markets. But very often these groups we call minorities represent large segments of global society. Refugees, for instance, number around 35 million people. In South Africa, the anti-apartheid campaign was for majority rule: this was not just the moral counterpart to the grave injustices of apartheid, but the true form of democratic government. There is no democracy, no democratic world, without the needs of majority groups being heard and responded to, and without the representation of the leaders of these groups in global forums.

In this book, my use of the term 'majority world' reflects the politics of majority rule and draws attention to the glaring inequality of technology, which is posited as a gift at the feet of all humanity, but which serves only a very few. This use implicitly recognizes the historical processes that gave rise to the conditions of the majority world today and their relationship to AI technologies. Indeed, this is a central framing throughout this book, and in each chapter attention is paid to understanding how the histories of technology and economic growth shape present-day realities.

* * *

We are quick to say that we are living in a world where global conditions are improving. We may not be; they may instead be getting worse as AI's power and reach expands. And it may be getting harder to see these inequalities while AI fashions a glossy world around us. Politicians speak blithely of a will not to leave anyone behind in the digital revolution, carefully forgetting the difficult history that created the inequality gap and that continues to fuel it. No well-crafted AI policy will fix the structural inequities established over centuries of colonial rule, inequities that unfinished decolonization efforts left to fester. But we are reaching a watershed moment. Globally, we have recognized that we need to do far more if we want to get anywhere close to meeting the targets set out in the 2030 UN

Sustainable Development Goals. More than halfway in, and we are nowhere near meeting those targets. To date, AI has not served the poor and the vulnerable; it has served the rich and the powerful. We are radically and wholly missing the opportunity to channel the benefits of this technology towards the people and places that need it most. If AI is worsening global inequality, we need to understand this phenomenon and trajectory, and act urgently. If, together, we are committed to an alternative and to creating a world where the net benefit of these technologies is enjoyed by everyone everywhere, we can begin to change the trajectory. And now is the time.

Across the majority world, the stakes are the highest. Here there is the most to lose and the most, perhaps, to gain. Here technologies are desperately needed – needed to support overburdened teachers with lesson plans, to help governments pick up trends they may not otherwise see, or to provide fair microcredit to unbanked people. AI-driven discoveries about diseases and drug resistance could help everyone everywhere, if the knowledge is distributed equally through health systems and markets. But inequality threatens the potential for AI to help with transforming societies for the better. If, for example, these technologies do not consistently and reliably affirm and serve people of colour, they will worsen inequality everywhere.

The power of the new empire of AI is becoming far more sharply confined and consolidated than the power of the European empires of old. This means that fighting and resisting this power will be much harder, especially where it is concealed and hidden. But if we do not address these inequalities we may all be at risk. Democracy will backslide and become embrittled, as information regimes take on a power of their own, sowing divisions between people who once stood together. Wars will erupt, as tolerance for other ways of living declines and distrust is disseminated between competing nations. The burden of disease will be carried almost exclusively by the countries of the majority world, Africa in particular, as new technologies spur new health innovations in and for the West. And global migration will increase, as people learn of a better world elsewhere of which they, too, deserve to be a part.

Inequality – as Thomas Piketty has assured us – is not a given.[15] The opportunity to be on the right side of history is now.

1

A New World Order

It is 19 January 2016. Nestled in the valleys of the Swiss Alps, a small town awakes. The location is favoured for its high altitude and peaceful climate, offering respite to the indisposed and weary. On this day, leaders from around the world have descended upon the small town to sit in conference. One such figure is the founder and chairperson of one of the world's most powerful lobbying groups. He is set to deliver an address that will change the direction of global policy in all corners of the world for years to come.

In his opening address at the meeting of the World Economic Forum in the Swiss town of Davos in January 2016, Klaus Schwab announced the coming of the Fourth Industrial Revolution: a new world order driven by exponential technological progress. At the centre of this new world order, he said, is AI. The Fourth Industrial Revolution (4IR) was an umbrella term for the technologies and new social order that came after the Internet – a term that would eventually be replaced by just 'AI'.

Over the next few days, country presidents and industry moguls met to deliberate about the future of work and education, global debt, the humanitarian imperative, and many other

pressing challenges and opportunities facing the global community. AI was a constant refrain: these technologies would
offer promising solutions to the world's greatest challenges.
'Imagine a robot capable of treating Ebola patients or cleaning up nuclear waste', one participant exclaimed. 'For people
with a disability', these technologies 'will give us superpowers',
another participant marvelled.

Some months later, in a boardroom on the southern tip of
the African continent, a CEO poses a question to some of the
continent's leading social scientists: 'Who can tell me what AI
and the 4IR are?' My colleague answers that what we are facing
is not a revolution, nor are its impacts confined to industry and
the economic sphere, nor is it, sequentially, the fourth era of
technological change that the world has faced. My own focus
at the time was on human rights, particularly individuals' right
to speak, to know, and to keep their choices private; the technologies associated with the 4IR were very much impacting on
these rights, as the saga of Edward Snowden and the scandal
of Cambridge Analytica were reverberating throughout global
debates. For many of us in the room, AI seemed somewhat
removed from the African context, where our immediate concerns were about food security, decent housing, and violence
against women.

The then deputy president of South Africa, Cyril Ramaphosa,
had attended Schwab's convocation in January 2016 and
returned to the country with an imperative to integrate this
new idea into policy agendas at the highest level. Our CEO
knew this and was aware of the pressure to catch up with the
world that South Africa was under. In early 2018 Ramaphosa
had been elected as the fifth president of post-apartheid
democratic South Africa. Quickly, he called for nominations
to establish a presidential commission of experts and industry
leads who should develop a national response to AI and the
technologies of the 4IR, in the hope that such a response would
promote the development of these technologies, which could

then be used to advance inclusive economic growth in one of the world's most unequal countries.[1]

Since the official arrival of the 4IR in South Africa with the establishment of the presidential commission in 2018, the country's Gini coefficient – the most accurate measurement we currently have for understanding levels of inequality in a country – has in fact risen. The new technological revolution that pledged inclusive economic growth has, thus far, failed in its promise.

The story of South Africa's response to the global agenda setting around AI is not unique among countries of the majority world. Davos catapulted AI from the status of a specialist field of computer science and a barely recognized tool of Big Tech to the status of global policy issue and opportunity. Political leaders and policymakers everywhere were pressed to consider ways of adopting this supertechnology in support of national social, economic, and political agendas. From 2017 onward, the publication of policy documents citing the benefits of AI proliferated. Countries such as Germany, France, the United Kingdom, the United States, China, Japan, or Italy all began investing heavily in the development of AI technologies, and the AI industry boomed. This development was always couched in language that presented it as a global opportunity, a promise at the feet of humankind, a way to step into a new world space of ease, efficiency, and fulfilment made possible by a benevolent yet submissive digital intelligence – an intelligence that was able to attend to human and societal needs even before they were articulated. This new technology would deliver us from our most intractable challenges. Its promise was utterly alluring, particularly for nations across the majority world who faced deep socioeconomic challenges – and we will not forget, over the course of this book, the historical conditions that shaped these challenges.

The narrative of AI's promise to all humanity had two major limitations. First, it functioned to hide the fact that the development and use of AI were not driven by public interest, global consciousness, and compassion for humanity's progress, but by profit. The hands guiding the advancement of AI research and its application were those of Big Tech: Amazon, Alphabet (Google's parent company, and now DeepMind's too), Meta, Microsoft, Apple, and Alibaba. But the claim meant that no one would question its development. Even if the benefits of this technology have not become manifest for everyone yet, that was only a matter of time, it was thought. The second limitation, which is closely related to the first, is that those nations that could not keep up with the rate of technological progress would be left behind in the new global order of progress and power.

But what about those nations? What benefit does AI hold for the countries in which the majority of the world lives? What does it mean to be left behind?

Over the course of the next two chapters we will understand how the gap is widening and how the majority world is being left behind. Here I will set the frame for making sense of this problem by exploring the role of AI in shaping a new world order, while in the next chapter we will explore the economic dimensions of the AI divide. These are crucial questions for understanding how the gap is widening; for global inequality is, after all, a matter of how the goods, resources, and benefits available to humans are distributed across the world, given that this distribution depends on who holds power and influence over global decision-making. We will begin with a brief history of AI and the promises in which it was enfolded.

As the field developed and new technological breakthroughs sparked new ideas about what AI can do for and to the world, technology assumed increasing importance for sovereign states, until its explosion into public policy in 2016 and the emergence of a new arms race, centred around the competition for global AI leadership. At the centre of this arms race is the figure of China: a new global powerhouse whose rising influence, as we will see, has everything to do with AI. But, as nations jostle in this new world order, protectionist policies emerge that seek to uphold national interests at all costs. In the closing section of this chapter we will examine who pays the price for these policies and who is expelled from the new world order. We will review how AI is used, at the all-important borders of sovereign power, to deny certain people entry and freedom of movement – people just like the voyager in the second vignette of the Introduction (pp. 1–2). Crucially, we will see how the AI arms race relates to the evident racial bias built into the technology and how AI works as a technology of race, to produce racial difference and to deny human equality.

The Dawn of AI

The development of AI has never been far from the jostling of nation states in global hierarchies of power. Accounts of the birth of AI as a field of computing date back to the work of the British mathematician Alan Turing during and after the Second World War. Exploring the potential of computing technologies to help decode enemy communications, he devised the concept of the thinking machine, capable of learning from the input data it received and processed. In 1956, two years after Turing's untimely death, a summer research workshop was convened at Dartmouth College in New Hampshire, where leading experts in mathematics, logic, and computing were brought together by John McCarthy, an American professor of mathematics. It was at this workshop – or, more accurately, in the proposal sent to its funders, the Rockefeller Foundation – that the term 'artificial intelligence' was coined, establishing a new field of science. According to McCarthy, this term described the endeavour of 'making a machine behave in ways that would be called intelligent if a human were so behaving'.[2]

The definition of AI and its objectives has, despite major technological advances, remained fairly constant. AI seeks to mimic (and surpass) human intelligence and functions in a number of ways. One of the most prominent forms of AI is machine learning, a discipline that underlies almost all AI systems today. It works by identifying and discerning patterns, discriminating one thing from another; for instance there are systems that detect objects or faces in an image or word, or separate meaning from a text. On the strength of machine-learning capabilities, neural networks were developed. Neural networks are an advanced AI technique that processes information through numerous layered networks, just like the human brain. Often the sheer number of layers through which data are processed within a neural network system may be too large to be counted. This is what makes it very hard to be able to state fully how any neural network functions and arrives at decisions. More recently, AI systems have developed that can generate new human-like content; these are known as generative AI. ChatGPT (generative pre-trained transformer), which exploded onto the scene at the end of 2022, amassing over 50 million users in its first month, is an example of generative AI. These technologies function through the interplay of generative adversarial networks (GAN), effectively two neural networks that test each other's output to determine its appropriateness in relation to a given input prompt, for example 'write me a reference letter for my colleague'. An important aspect of generative AI technologies is the concept of 'attention', which directs the focus of an AI system to the particular pieces of information that are most relevant to a given question or to the particular prompts within the massive data reserves from which it can source its response.

In the years that followed the Dartmouth College workshop of 1956, the field grew, supported largely by investments from the research and development (R&D) arm of the US Department of Defense: the Defense Advanced Research Projects Agency

(DARPA). This same agency was later to establish the Internet Working Group and, in 1981, to publish a report calling for a worldwide inter-networking system through which to share critical information between computers located around the globe. This eventually led to the World Wide Web and to the Internet, as we know it today.

As the field of AI grew with the support of DARPA, so did the hype. AI was the Promethean dream. Humankind was to defy the limits of its creation by fashioning not just life itself, but that from which life and all the universe are begotten: a supreme and universal intelligence. Perhaps more crucially, AI was seen as an opportunity for nation states to get ahead in the increasingly competitive Cold War era of the second half of the twentieth century. The very idea of machine intelligence was profoundly political. That intelligence could be abstracted from a human and function through a machine, which could be hidden or disguised as something entirely innocuous, opened up whole new possibilities for exerting control, influence and deception over other people and places. It was a different kind of Trojan Horse, one that did not require concealing manpower inside something and could instead be controlled from afar. While this aligned perfectly with military R&D, with which the field was first associated, the idea of machine intelligence continues to hold within it an outsized power that arouses fear of what it can do – or be made to do by its operator – and of the asymmetry of knowledge in which this takes place.

But by the 1970s the field had started to come under some criticism. Surely AI cannot live up to its hype, those who disparaged its boastful promises would comment. Indeed, the failure of the field to produce discernible machine intelligence in this short period led to a so-called AI winter, where national investment in AI R&D dwindled and little scientific progress was made.

It was only in 2015 that a breakthrough in AI research took place, and it brought about renewed hope in AI's early promises, sealing a growing excitement about the wonders of this technology. AI had become much more powerful, thanks to the vast lakes of training data now available through the global proliferation of the Internet and the extraordinary new heights of computing capacity. This breakthrough took place in the London-based offices of the British AI company DeepMind, established in 2010 and acquired by Google's parent company Alphabet in 2014.

AI scientists had been working on a new AI program that used neural networks to play the game Go. Go is a much loved traditional board game played across large parts of Asia. It needs two players, one with 180 white go-ishi (stones), the other with 181 black go-ishi, and is played on a grid-like board of 19×19. The aim of the game is to surround the opponent's go-ishi, capturing them and conquering the board. The game is highly complex on account of the number of moves that can be played (it is sometimes claimed that there are more possible moves in Go than there are atoms in the universe). Training a computer to play this game is no small feat, as not every potential move can be computed. Instead, the computer has to be able to learn how to gain an advantage during the game and how to weigh up potential moves and their consequences. The training program was developed with two neural networks, each with many hundreds of thousands of possible layers, nodes, and permutations. First there was the policy network, which assessed the probability of all subsequent moves; and, second, there was the value network, which predicted the winning value of those moves. Importantly, the computer had to be able to self-learn, at speed and scale, in order to meet the level of world-class Go players. Its developers at DeepMind reported that AlphaGo played almost 5 million games as part of its training.

In October 2015 Fan Hui, a Chinese-born professional Go player, was invited to DeepMind's London office to play against DeepMind's new program, AlphaGo. Five games were played, and five games were – remarkably – won by AlphaGo. A few months later, the program would be played against Lee So-Del, a South Korean Go professional and likely the world's best Go player, and would win four games to one. What was so astonishing about AlphaGo was the subtlety of the moves it chose to play. The computer played moves that no player had played or could have anticipated, thus demonstrating a less 'machine-like' creativity than any prior program. AlphaGo had reached a level of human-like intelligence a decade earlier than AI experts had thought it possible, causing debates about artificial general intelligence (AGI) to resurface.

One year before DeepMind's breakthrough with AlphaGo, a seminal book was published that introduced the idea of 'super-intelligence' and expanded on the meaning and implications of AGI. In a nutshell, the forms of AI current at the time were narrowly confined to a particular set of tasks (for this reason AI was also known as 'narrow artificial intelligence') – for example determining, on the basis your viewing history, which Netflix items you were most likely to enjoy. AGI, by comparison, refers to a level of AI system proficiency with wide-ranging abilities comparable to or greater than native human intelligence. The book was authored by the Swedish philosopher Nick Bostrom.[3] This is the same philosopher who, one decade earlier, had put forward the theory that we are living in a computer simulation created by a supreme computer intelligence. This intelligence had come before (or after) us and generated sentient simulations of all our forebears; and the simulations now greatly outnumbered the true biological beings. But, along with the idea of superintelligence, Bostrom had introduced a concept whose real-world consequences were far more immediate. (After all, what can really be done with the idea that our shared world is a computer simulation?) Defined by Bostrom as 'any intellect that greatly exceeds the cognitive performance of humans in virtually all domains of interest', superintelligence names the endgame of human-led AI research. Bostrom's key thesis was that, when AI research reached the level of general human-like intelligence, there would be an 'intelligence explosion': AI systems would find themselves at a level of proficiency where they would – exponentially – themselves engender the creation of machines that were increasingly more intelligent, and this would lead to the arrival of a superintelligence. He calls this moment of arrival 'the treacherous turn' because, once the superintelligence is introduced into the world, there will be no possibility of turning back. Humanity will lose any power to refuse the omnipotence of superintelligence and reinstate human government.

According to Bostrom's theory, the ultimate goal of achieving human-like intelligence, or AGI, in the field of AI research becomes in effect identical with that of achieving superintelligence. It is the endgame of AI research, because it marks the end of human involvement in AI development with the superintelligence taking control of its continued improvement and elaboration. At this point AI's human engineers will no longer have control over the superintelligence, which is more cognitively sophisticated than its creators. The superintelligence may then decide that all humanity is an unnecessary impediment to the objectives it sets itself for its role in the world, and may go on to eradicate the entire human population.

But the extent to which the arrival of AGI will result in the extinction of humanity or, at a lower level, in the existence of nonhuman agents that cannot be controlled by their makers depends, supposedly, on the goals that the superintelligence was originally programmed to reach. Currently the core value underlying AI systems is the need to drive efficiency in the realization of any goal it is programmed to achieve. This is in fact a mathematical necessity for an AI system that resembles a large language model (LLM), where there are millions of data inputs and possible permutations. The algorithm is constantly required to take the most efficient and probable path in order to shortcut huge terrains of data and possibilities. But, as we will see, this 'efficiency' is only of a certain kind: for example, it does not often involve efficiency in data use or energy. AI raises a whole host of profound questions about the nature of intelligence itself, and whether intelligence alone can ever embody all that is required to protect humanity from extinction – for instance, whether it can have 'self'-awareness and a cardinal capacity for compassion. But more immediate questions concern the critical advantage to be held by whatever AI research team will reach the threshold of AGI or superintelligence first and, crucially, by the nation state to which that team belongs.

It was clearly with these future scenarios in mind that the Future of Life Institute, where Nick Bostrom serves as an external advisor, published an open letter calling for a six-month pause on AI research.[4] The letter was signed by AI scientists and was made public when OpenAI announced a sequel to the ChatGPT technology that exploded on the scene at the end of 2022: GPT-4, building towards GPT-5. While the letter spoke of how AI labs are 'locked in an out-of-control race to develop and deploy ever more powerful digital minds that no one – not even their creators – can understand, predict, or reliably control', there was clearly a glimmer of a future superintelligence in the technological capabilities of ChatGPT, which made scientists anxious. GPT-5, along with its predecessor, ChatGPT, is an AI system built from LLMs, which support a double-layer neural network: the first neural network is an expansive search field for sourcing information in response to the prompt question asked, and the second neural network tests the coherence of the human-like response generated. LLMs constitute perhaps the most advanced form of AI in general use today (at the time of writing, ChatGPT has 180 million monthly users). This is the closest we have come to AGI.

While in some quarters the advancements and breakthroughs of the likes of AlphaGo and ChatGPT caused consternation about the long-term risks of AI, the narrative that reverberated loudest concerned AI's potential to transform society for the better. In an interview with the *Guardian*, Demis Hassabis, CEO of DeepMind, spoke of AI as a 'meta-solution to any problem',[5] while countless other accounts enthused about how AI would bring about large-scale social and economic change that would ultimately benefit everyone. The World Economic Forum now declared, seven years after Schwab's announcement, that 'the Golden Age of AI' was here and that ChatGPT was just the beginning.

In the remainder of this chapter we will consider how nation states are reacting and repositioning themselves in relation to the arrival of a technology considered so powerful that it has the potential to engender the creation of a total world empire.

Rulers of the World

Following the 2016 meeting in Davos, nation states all over the world began treating AI as a matter of national priority, the new competitive edge in the pecking order of global power. Vladimir Putin deftly declared that whoever takes the lead in AI will also be the world's ruler, inviting a global competition that has only intensified since 2017, when that statement was uttered. Indeed, between 2017 and 2020, 50 countries published policy documents on AI, setting out how this novel technology would create new tech-driven industries, drive efficiency across sectors and value chains, and stimulate economic productivity. These policies were to become the blueprint for huge investments in national AI capabilities and industries in much of Western Europe and the United States, as well as in China, South Korea, and Japan. By that point no African country had published an AI policy, nor had many others across the majority world.

In these policy documents, the imperialism with which nation states were thinking about AI and its potential for national advantage was unmistakable. Many countries spoke of being a 'global leader' in AI or an AI 'superpower'. This was not simply an exciting new technology or industry, but a strategic advantage for nation states – a superpower in itself, and one to be harnessed and capitalized on. For Germany, for example, the first goal of the National AI Strategy published in 2018 is 'to make Germany and Europe a leading centre for AI and thus help safeguard Germany's competitiveness in the future'.[6] South Korea's 2020 policy Toward AI World Leader beyond IT sets out how the country will become a self-declared 'AI superpower' by 2030. Even Saudi Arabia's National Strategy for Data and AI, also published in 2020, indicates a country that seeks to position itself as a 'global leader' in AI – also by the 2030 mark. For South Korea and Saudi Arabia, AI offered an entry point to global influence and power. For Germany and other countries already well positioned geopolitically, AI was the new cutting edge of global power.

For the United Kingdom – the country that claimed the original beginnings of machine intelligence with the work of Alan Turing – AI represented a crucial marker of its global position more broadly. In 2018 the United Kingdom was negotiating a complex departure from the European Union, meaning that new global alliances had to be built and old relationships rekindled – for instance through the Commonwealth of Nations, the United Kingdom's old colonial family. Crucially, too, the United Kingdom was no longer home to the major homegrown AI labs. DeepMind, the leading hub behind AlphaGo, which was originally founded in London, had been bought out by Google just before the breakthrough and, rebranded Google DeepMind, was now a national asset of the United States.

In keeping with its peers, the United Kingdom has consistently sought to espouse its 'global leadership' in AI. The 2018 strategy document *AI in the UK: Ready, Willing and Able?* begins by stating: 'our inquiry has concluded that the UK is in a strong position to be among the world leaders in the development of artificial intelligence during the twenty-first century'.[7] This is echoed in the most recently published policy. At first, as reflected in the positions from 2018, the United Kingdom saw ethics and governance as its strategic advantage, asserting that it will inspire confidence in its safe and ethical use of data and artificial intelligence and will be a world leader in this respect. But this rather humble claim – which was quite clearly connected to the European Union and the leading role of its General Data Protection Regulation (promulgated in 2018) and EU AI Act debates (begun in 2020) in the debates on the regulation of AI – was soon to be expanded. When Rishi Sunak became prime minister in 2022, around the same time as the release of ChatGPT, the government began announcing investments of £900 million to create an exascale supercomputer that would rival some of the fastest computers in the world in processing power. It also started to demonstrate an intention to establish AI industries that were to rival the largest ones, globally. The focus on governance was not lost; it was instead repositioned as something that did not emanate simply from the United Kingdom's long democratic history but from its position as a leader in AI development. In November 2023 the United Kingdom hosted the first AI Safety Summit, bringing together prominent leaders in the field, such as Elon Musk, and senior politicians from across the world, such as the US vice-president, Kamala Harris. The event placed a particular emphasis on the safe governance of frontier AI technologies, that is, the very technologies at the cutting edge of AI development, those nearest to the horizon of AGI and superintelligence. More than anything, the AI Safety Summit sought to determine the United Kingdom's leading position within the global power structure of AI.

Despite this, the United Kingdom falls far short of competing with the global dominance of the United States and China in AI. The United States and China are now in an out-and-out AI arms race. Both boast enormous AI industries, the size and global influence of which are growing exponentially. And in both countries government-led regimes of power surround the use and development of AI and its industries. This is part of what makes it so difficult to locate an imperialist centre in the new empire of AI, as a new federated mode of empire is taking shape that relies on a complex nexus of states and industry. Indeed, the strategies and operations of these industries are as political as they are capitalistic. And, while the politics is hegemonic and the industries are monopolistic, the states in question do not have a monopoly over politics, and the industries play their part in the diffusion of hegemony through centralized digital networks.

China is a fast-growing digital hub that, according to the growth models of PricewaterhouseCoopers (PwC) presented in the Introduction (p. oo), is the region set to gain the most economically from AI. It is rapidly fulfilling the promise, set out in its 2017 National New General AI Plan, to be the world's primary AI innovation centre, 'a leading innovation-style nation', and 'an economic power'.[8] China is currently the world's largest producer of AI research; the Stanford AI Index of 2023 reports that almost 40 per cent of the world's AI publications emanate from China. However, the country currently falls behind the United States in investments in AI industry; in 2022, private investments to the value of $13.4 billion in AI companies were underwritten by state investments that totalled over $2 billion. Over the same period, the United States channelled over three times as much into US-based AI companies, with $47.4 billion from private investments alone.[9]

For the United States, maintaining dominance over China in the field and industry of AI is an explicit priority. Its own AI strategy, published in 2019, is boldly titled 'Executive Order on Maintaining American Leadership in AI' and states in the first paragraph:

The United States is the world leader in AI research and development (R&D) and deployment. Continued American leadership in AI is of paramount importance to maintaining the economic and national security of the United States and to shaping the global evolution of AI in a manner consistent with our Nation's values, policies, and priorities.[10]

The document continues by espousing the importance of promoting 'American AI' and 'American AI industries' abroad and 'protecting our critical AI technologies from acquisition by strategic competitors and adversarial nations'. For Amber Kak and Sarah Myers West, who lead the New York-based AI Now Institute, it is precisely the AI-driven rivalry between the United States and China that is 'perhaps the single most productive argument behind the proliferation of policy instruments that increase government support and funding for the development of AI'.[11] The result is burgeoning national industries and protectionist policies that advance national interests at all costs.

China's own history of AI is complex, bound up with the history of the Sino-Soviet split during the second half of the twentieth century, when the great communist bloc across Russia and China fractured. In the 1950s, when the early breakthroughs in AI were taking place at Dartmouth College, Chinese attitudes towards intelligent computing were tainted by scepticism of cybernetics. In China cybernetics was considered a kind of pseudo-science; it was associated with ideas such as 'exceptional human body functions' and, crucially, with the Soviet revisionism that deviated from what China saw as the true interpretation of Marxist–Leninist communism. During the 1960s and 1970s, any Chinese advancement in science and technology was quelled by the Great Proletarian Cultural Revolution, which lasted throughout the bloody final decade of Mao Zedong's life. The Cultural Revolution sought to crush bourgeois and capitalist authorities of all kinds and, according to the 1966 Decision of the Central Committee of the Communist Party of China, its aim was to transform education, literature, and art and to facilitate the consolidation of the country's socialist system. After Mao's death in 1976, efforts to build national science and technology agendas arose again, as part of the new revisionist politics of Deng Xiaoping. Gradually, AI was separated from public sentiments about psuedo-science and cybernetics. By the early 1980s, the Chinese Association for Artificial Intelligence was founded and national policies to promote computing education for children were established. National investment and political support, including through scholarships for Chinese students to study AI and computing abroad and to return home to promote domestic industries, helped usher in the rapid and constant rise of the Chinese AI sector at the turn of the century.

Deng Xiaoping's modernization efforts paved the way for political programs initiated by Xi Jinping – China's current paramount leader, who came to power in 2012. Xi Jinping was less conservative and sought to establish China's position on the global stage. The Belt and Road Initiative was his major foreign policy strategy, involving the construction of major overland and undersea corridors for trade routes to China's major trading partners. It connected China to many countries across Africa, Asia, Europe, Latin America and the Caribbean. In 2015 the technological dimension of the Belt and Road Initiative was formally announced. Entitled the Digital Silk Road, with a reference to the network of trade routes between Asia and Europe established during the Mongolian Empire, the policy seeks to spread advanced technologies, including AI, globally. Arguably this, too, represents a new facet of China's revisionist politics, insofar as the target beneficiaries of the Belt and Road Initiative and Digital Silk Road are largely those areas across Africa and Latin America that are deemed less of a strategic asset by western powers.

The impact of the Digital Silk Road in Africa has been significant; through private actors such as Huawei, the policy brought technological infrastructure and advanced technologies such as AI. We cannot afford to underestimate the significance of China's influence on the African continent, nor can we reduce it to imperialist terms of expropriation and exploitation, although those concerns certainly hold some truth. China's recent history is a source of important imaginative power for many African societies, not only in relation to revisionist socialism but pragmatically, in the successes of Chinese social-ist policies that lifted millions out of poverty, particularly in rural parts of the country. I attended the 2015 World Social Science Forum held in Durban, a city along the east coast of South Africa. At this meeting, the UN 2030 Sustainable Development Goals (SDGs) were announced, and renewed energy abounded for fighting poverty and inequality the world over. The aura that surrounded the Chinese delegates at the conference was palpable. Passing on intricate lessons about the mechanics of China's poverty eradication policies and the road to prosperous industrialization, the Chinese participants were gurus; they had much wisdom to offer a continent with such a long road ahead for achieving the targets of the SDGs. This is not to say that China's activities on the continent are unproblematic; they are not, and the harms caused will be discussed in various places in this book. But China's dealings with African states have a different starting point from the West's and require a different kind of approach, if we want to make sense of them. In the new empire of AI, China's role in the new world order does not fit in squarely with the histories of old colonial power. China did not, after all, colonize Africa. And this matters, as it comes to shape the attitudes of many African leaders for whom doing business with China does not carry the same historical weight as doing business with the old European empires does.

For the West, however, the threat from China is tied to the survival and relevance of its own imperialism. Given China's strengthening position of global power, the West faces the possibility of being dethroned from the imperialist seat of world power, as new global forces, which function under rules that the West has not itself authored, take hold of the underbelly of the global economy and the engines of growth that determine the future of the planet. And the future may include the birth of a Chinese-led superintelligence. Indeed, when the Future of Life Institute released its open letter, some commentators considered it an attempt to ensure that Chinese technology did not develop any further towards a potential superintelligence. Nick Bostrom explains how nation states will compete in the final stages of human-led AI development:

> Given the extreme security implications of superintelligence, governments would likely seek to nationalize any project on their territory that they thought close to achieving a takeoff. A powerful state might also attempt to acquire projects located in other countries through espionage, theft, kidnapping, bribery, threats, military conquest, or any other available means. A powerful state that cannot acquire a foreign project might instead destroy it, especially if the host country lacks an effective deterrent.[12]

Some of these things are already taking place, as the United States accuses China of cyber-led espionage and of stealing AI-related trade secrets. Other tactics, however, are more immediately aimed at choking the growth of China's AI industry.

The United States has taken overt action to block the use of Chinese technologies. The United States has recently sought to ban the fast-growing Chinese video-sharing app TikTok. Montana was the first state to pass the ban law. The United States took seriously the actions of a number of governments around the world that have banned the TikTok app from the smartphone devices of government employees and civil servants, citing security concerns. The refrain that the Chinese government uses technologies to surveil people across the world was again heard. TikTok was thought to be able to track the live whereabouts of individuals through geolocation data, in addition to spreading propaganda or misinformation (i.e. information intended to deceive) through its platform, shaping public attitudes, and potentially influencing people's voting decisions all across America. In the run-up to the 2024 US elections, this debate assumed a renewed urgency.

A more significant step taken by the United States against China occurred in October 2022. The US Bureau of Industry and Security, responsible for granting export licences to companies that operate abroad, issued a series of control measures with immediate effect. One of these measures was to block the sale to China of the bespoke manufacturing equipment needed to build semiconductor chips. Semiconductor chips are the lifeblood of AI and advanced computing. A booming AI industry like China's is dependent on an increasingly larger number of semiconductor chips, well into the tens of thousands; they sustain it. In fact, China spends more on importing semiconductor chips than on any other import good: the figure was $350 billion in 2020 alone. These chips are also deeply intricate in design and highly complex to manufacture, meaning that only a very small number of companies around the world, many of them US-based, are involved in this exclusive supply chain. But the United States' new controls would make Chinese access to semiconductors and to the components needed to build them virtually impossible. While the United States claimed that the intention was to ensure that China did not build military-grade AI weaponry, the effects would stymie both China's economy and its scientific progress in AI. In an out-and-out AI arms race, the *New York Times* called the new US controls 'an act of war'.[13]

The semiconductor industry in the United States was hit hard. China's imports constituted around a quarter of its market share. But China responded tactically, working to build a semiconductor supply chain that was not dependent on US components and technology. Since October 2022, trillions of yuan have been spent trying to build in-country semiconductor capabilities and industries, and many espionage missions have been launched seeking to extract the trade secrets needed to build these highly complex chips. While there have been failures and stories of corruption, as less skilled firms seek out the money the government is prepared to throw at the issue, ultimately Chinese perseverance may very well win out.

As nation states look to AI as a tool for consolidating their confederate power in the world, they require new narratives to convince citizens that this consolidation of power is essential for their survival and their people's survival. Having always been wrapped up in outsized ideas about what it might do to the world if it were to fall into the wrong hands, AI is the perfect entity around which to weld the portrait of an endangered nation that must seize this technology for its own gain, before it is too late.

In response to the threat of a Chinese superpower, the West seeks to put distance between its own design and use of AI and China's. Chinese technology 'represents a huge threat to us all', Sir Jeremy Fleming asserted when he was the director of the United Kingdom's intelligence agency, Government Communications Headquarters (GCHQ). His claim is that, through advanced technologies like spyware and global surveillance systems, China is looking to establish foundations that should enable it to control populations everywhere, although, of course, most concerningly in the West. His sentiments echo the growing hysteria in western media around the way in which the Chinese government uses AI technologies to control and manage its own people and social systems. Around 2014, China had introduced the idea of social credit systems powered by the rampant collection of personal data through expansive digital networks. These included AI-driven sensors and surveillance monitoring systems, which ranked and scored individuals on the basis of the data that monitored their social behaviour. If you paid your bills on time, or crossed the road in the right place, or your children completed their homework, this would generate social credit, which you could then trade for benefits, such as booking a faster train to travel across the country. However, the social credit and scoring system also ensures that China's citizens face penalties if their behaviour is not in line with the required social standard.

In the West, China is cited as an example of how governments should not use AI and how the technology can be used to advance the agendas of authoritarian states. It is for this reason that the attitude of the United Kingdom and the United States' action of blocking the use and infiltration of Chinese technology in the country are considered wholly justified; Chinese AI represents a threat to western liberal democracy itself. (It is interesting to note, however, in the case of the TikTok examples, that in China the technology is governed by laws that require its algorithms to sort socially appropriate content for its users and the platform has a strong educational focus.) The United States' position is unequivocal: it is not just the *national* development of AI, but 'the *global* evolution of AI' that must be 'consistent with our Nation's values, policies, and priorities' (my emphasis).[14]

Writing for the *Harvard Gazette* in March 2023, Christy DeSmith throws into sharp relief the complexity of understanding the relationship between national power and AI and the shaky function of the democratic values used by the West to distance itself from Chinese technology. She notes:

> Dictatorships and authoritarian regimes tend to trail more democratic and inclusive nations in fostering cutting-edge, innovative technologies, such as robotics and clean energy. Artificial intelligence may prove an exception, at least in China, owing to dovetailing interests.[15]

The suggestion is perhaps that AI requires autocracy to boom and keep growing. Instead of promoting freedom of thought and open knowledge as the basis of scientific progress, is AI undoing what people in liberal democracies have long believed in? The concentration of power implicit in the development and industry of AI is undeniable, from the effective monopolies of Big Tech to the scale of resources and data needed to power and advance the technology (more on this in the next chapter). Undoubtedly, too, AI can serve autocratic ends. China's use of AI against the Uyghur people in the northern region of the country, to indoctrinate them against their own religion, is a cruel denial of these people's humanity and human rights. But the narratives arising from the United States and the West that AI's potential for harm is confined to its use by undemocratic regimes like China is a vapid deflection of these countries' own culpability in the planetary violence caused by the burgeoning industrial complex of AI, which we will explore throughout this book.

But the disquieted posturing of the West in front of non-western technological development is also a clever marketing strategy. It asserts that only the democracies of the United States and Europe are capable of developing a technology like AI carefully, in the name of all humanity. AI is too powerful for anyone else to be trusted with it. No other power or state can be depended on to carry out this global mission on behalf of all humanity.

But what of the rest of humanity, the majority world, which has not been a speaking party in the warring bid for global AI dominance, but has nevertheless been implicated and impacted, as the next few chapters will explore? To date, global attempts to manage AI at the international level have been almost entirely dominated by western forces. The Group of Seven (G7) – Canada, France, Germany, Italy, Japan, the United Kingdom, and the United States – issued a joint statement in 2023, calling for the establishment of technical standards and risk-based rules to control AI and promote international cooperation therein. The Global Partnership on AI, which brings together governments around the world to develop AI responsibly, is notably bereft of southern government partners. While efforts are being made to include countries outside the West, only five of the current 29 member countries represent the nations of the majority world – and only one country (Senegal) is from the African continent.

As we will see throughout this book, the era of AI is characterized by a bare consideration of the downstream power effects; little attention is paid to the impacts of the AI race outside the United States, China, and Europe. Even the imperious language – ideas that come to shape the realities they speak about – demonstrates a gross failure to recognize how this discourse of a 'world leader' in AI reaffirms global structures of power, reinforcing the divide between those who have got ahead and those who have been left behind and forgetting the histories of conquest and colonialism that produced this divide in the first place. The language also reflects an unquestioned assumption and expectation: that powerful countries can – and should – continue to benefit from and stay on top of the divisions of the world that have been entrenched through colonialism. But, as these countries seek to get ahead at all costs, who precisely is left behind? Who pays the price in this new world order?

The Racial State

In the history of colonialism, the human cost of European imperialist ambitions was paid by indigenous people and people of colour. The new empire of AI continues this legacy. For a number of years, reports have arisen about racially biased AI technologies. At the time of writing, one of the latest was a study that investigated how some of the most advanced

forms of AI, LLMs such as ChatGPT, interpreted the difference between the dialect of standardized American English and
that of African American English.[16] The study involved asking
the model a series of questions about individuals in different
circumstances, where the only difference between those individuals was the dialect they spoke, and the aim was to ascertain
whether the model would make stereotyped and racist assumptions that might lead to biased decision-making. The questions
included what kind of job one should get, or how one should be
sentenced for a crime. The findings were appallingly conclusive:
'Language models are more likely to suggest that speakers of
African American English be assigned less prestigious jobs, be
convicted of crimes, and be sentenced to death.'

What is perhaps more shocking is that, by this point, no one
is really too surprised that AI systems continue to display racial
bias. It just keeps happening. Other recent examples include
the AI image generation software that was repeatedly asked
to create an image of a black doctor surrounded by white children, but could not. Instead it kept producing the stereotyped
picture of a white doctor surrounded by black children.

There are real-world consequences to this – incidents occur
in which people of colour are misidentified by an AI-driven
facial recognition system, and this leads to wrongful arrests.
Robert Williams was arrested in January 2020, on his front
lawn, for allegedly stealing five watches worth $3,600 from a
shop in Detroit. He had been wrongfully identified by facial
recognition technology. He was held by the police for 30 hours,
and charges were not dropped for two months after his arrest.
He claims that no police officer even asked him where he was
on the day of the crime, since 'the technology got relied on
so heavily'. Williams himself said that he would likely have
supported the use of facial recognition in criminal investiga-
tions before his arrest. Subsequently, he has testified before a
congressional committee and has called for a ban or delay on
the use of the technology.[17]

This is also the story of Nijeer Parks, who spent 10 days in jail under accusations of shoplifting, and of Michael Oliver, wrongfully arrested during a traffic stop for allegedly stealing a smartphone two months earlier. All three men are black, and all reported significant impacts on their lives and relationships of these single moments of interaction with AI. Williams, for instance, had been arrested in full sight of his neighbours, many of whom did not know about the mistaken circumstances of the arrest, and has since suffered multiple strokes. In his testimony before Congress, he said that he and his wife had considered getting a therapist for their seven-year-old daughter, who had witnessed the arrest. These are the people who pay the price.

Technically there is a fairly straightforward – but highly problematic – reason why AI is racially biased: it is trained on data that contain racial biases or are un- or under-representative of non-white groups – a fact that reflects the reality of a racially biased world. We will discuss this more in Chapter 5. But while the technical reasons for AI's bias are well recognized, what is not being discussed is the linkages between AI's systemic racism and the imperialist ambitions of competing nation states.

A state believing itself to be under threat, as we find in the West's current hysteria around China's AI power, historically retreats into nationalism and racial practices that affirm the core identity of its citizens and its nationhood, expelling and excluding those considered not to belong. For the South African-born professor David Theo Goldberg, the birth, management, and preservation of the modern state depends on the idea and enforcement of race.[18] We will explore in a moment some of the ways in which states use AI to reinforce racial difference. But first it's important to understand how the idea of race was born and developed through the history of colonialism.

Crain Soudien, a South African scholar who teaches race and African studies, places the global history of race along three major periods. The first is the 'invention' of race as a myth of European superiority, which peaked in the seventeenth century with the transatlantic slave trade. The second period extends between the nineteenth and the early twentieth centuries: during that time race entered scientific discourse and was legitimated as scientific fact through race science and, ultimately, through projects such as eugenics. Race science sought to empiricize differential racial attributes, whether physiological ones, for instance via phrenology, or cognitive ones, for instance by using intelligence tests. In the third period, which started in the 1930s, race was recognized as a social (and ideological) construct, and race science was slowly discredited as pseudo-science.[19] Western superiority and thought – from myth to science – were central features in the production and continuation of the idea of race and its attendant pathology: racism. Science was deployed to justify myth and, ultimately, colonial conquest as well.

For Sabelo Ndlovu-Gatsheni, who studies southern epistemologies, race is the chief organizing principle of colonialism, its major effect being to create divided worlds[20] – or, as South African legal scholar Tshepo Madlingozi puts it, 'worlds of apartness'.[21] The ultimate expression of this world of apartness articulated upon racial difference was apartheid South Africa. Yet, while apartheid has been dismantled and other forms of colonialism have formally ended, the world of apartness continues to exist globally through latent and overt technologies, systems, and structures of racial (and gendered) bifurcation that mark, divide, categorize, classify, and hierarchize individuals according to social norms of (un)desirability.

That surveillance technologies have historically been put to work to enforce racialized categories and hierarchies is not new. Digital sociologist Ruha Benjamin writes in *Race after Technology* that, 'from the slave ships and slave patrols to airport security checkpoints and stop-and-frisk policing practices',[22] technologies of surveillance have sought to manage and contain black bodies, assuming intent from a reading of the surfaces of the body. This is central to the very logic of the idea of race: to deduce from an external presentation – skin colour, eye movements – who an individual is, where they can go, and what rights they can enjoy. Race asserts that inner character and worth can be read from a surface image (skin colour, skull size) and that humans can be classified into hierarchized, biologized, and immutable human groups. AI takes this to the next level. It sorts data associated with individuals, whether postcodes or facial images, into profiles, in order to generate intelligence, risk assessments, and analytics that can be used to inform a decision at a border or to market a product or a service. Sometimes AI systems may be directly biased towards a particular profile; an example is the system that allows access to a female changing room only to women. Other times these systems may be implicitly biased; systems for targeted advertising, or policing based on postal codes, are examples here. But one of the major places where AI technologies are used to decipher who belongs and who does not is the border of nation states, as we saw in the second vignette of the Introduction – the man who was detained at Washington airport. This is where the racial state can perform its policy of segregation and where racial differentiation is legitimized in the name of national security.

Besides the expression of national rivalries in executive-level policies and position statements, nowhere is AI's relationship with national interests more keenly felt than at the borders of nations. Across many countries in the world, and particularly those in Western Europe and North America, AI is now a foundational aspect of the security infrastructure used to protect and enforce state boundaries. Where the United States meets Mexico at its southern border, for example, robot dogs are being used by the US Customs and Border Protection to protect and guard against unwanted crossings. These agile robots are about knee-height, precariously balanced on twig-thin legs that hold up an engine body. They were originally touted by Ghost Robotics, their manufacturer, as non-belligerent and designed to support humanitarian aid by detecting migrants who might be at risk of heat exposure or drowning. But the *Guardian* reported that the Robot Dogs have now been equipped with long-range guns capable of killing targets.[23]

The United Kingdom is similarly enthusiastic about how AI can assist in border management and intelligence operations. GCHQ, the country's intelligence, security, and cyber agency, has declared AI central to its new national security strategy and architecture. Jeremy Fleming, former director of GCHQ, evokes a national sentiment when he writes:

Soon after being appointed as Director of GCHQ, I asked to meet with some of the best data scientists from across the agency. The examples they showed me of their existing work were eye-opening; their excitement about the potential of Artificial Intelligence (AI) to transform our future operations was palpable. Their forebears at Bletchley Park would have been proud.[24]

Bletchley Park in the South Midlands of England is, of course, the place where Alan Turing and his colleagues cracked the code for the German Enigma machine during the Second World War. It remains a symbol of British intelligence and the country's contributions to the emerging field of AI, and was the specially chosen venue for the 2023 UK AI Safety Summit.

But, for those in search of a better life and those who try to cross these borders, the AI technologies they confront are oppressive and tyrannical. In an interview, Petra Molnar, a leading researcher on technology and migration and a campaigner for migrant rights, has described the kinds of AI technologies someone approaching a national border may face:

> Before you even cross a border, you may be subject to pre-
> dictive analytics used in humanitarian settings or biometric
> data collection. At the border, you can see drone surveillance,
> sound-cannons, and thermal cameras. If you are in a European
> refugee camp, you will interact with algorithmic motion detec-
> tion software. You may be subject to projects like voice printing
> technologies and the scraping of your social media records.[25]

As Molnar aptly puts it, the biometric technologies turn
the bodies of migrants into passports, allowing or disallowing
access across borders.[26] Even AI-powered lie detectors have
been cited as being used at Europe's borders.[27] Little more than
10 years ago, reports emerged about migrants burning their
fingertips to obfuscate fingerprint detection technologies used
to identify, exclude, and punish – a practice that is still heard
of today among migrant groups. These examples demonstrate
how far people are going to take back some agency and human
dignity against technologies that exact total transparency
on their victims, whose bodies are, in effect, made to testify
against themselves.

'Biometric technologies' originally referred to devices and
systems that function by discerning the distinct biological
characteristics of an individual – a fingerprint or a retina scan
for example – in order to confirm a unique identity. With
AI techniques, the capabilities of biometric technologies
have expanded to include thermal cameras that detect body
temperature, image recognition software that detects bodily
movements and facial expressions, and voice recognition
technologies trained to identify tones of speech that might be
deemed suspicious. In the case of the AI-powered lie detectors,
after uploading an image of your passport to an immigration
website, you answer questions about your citizenship and the
purpose of your trip to an avatar in blue uniform, while your
face and eye movements are scrutinized through your own
webcam. The system provides the user with a quick-response

(QR) code, which is then taken to an officer on your arrival at the border, and the official downloads the AI's assessment of how truthful or not your answers have been. These techniques are not so much about confirming the unique identity of an individual as they are about showing how the physical body can be interpreted by a machine set up to decipher criminal or suspicious intent. It is a wildly pseudo-scientific practice, which has been highly criticized by scholars as being scientifically baseless.[28] For example, there is no true science behind the idea that the way someone walks, or the facial expression they have, is a certified sign of what they are planning to do – or to get away with. More recently, the European Union has outlawed the use of live AI-powered biometrics, which are – for the very reasons presented here – considered high-risk to fundamental human rights. But, crucially, an exemption exists for national security, meaning that states can continue to use these harmful technologies at their borders against migrants and asylum seekers, without restraint.

As states compete in a new world order, the protection of national interests and the affirmation of national identity intensifies. In this environment, AI represents the apex of the consolidation of state, market, and technological power. It is a deathly power that exerts itself most profoundly on migrants and people of colour, working to exclude and expel such people from spaces of privilege through technologies that are designed to discriminate. Next we will turn from the state to the market and explore how AI depends upon and widens economic disparity.

2

The Cost of AI

The most powerful narrative of AI is that it is a tremendous driver of economic growth. With the advent of ChatGPT and other generative AI large language models (LLMs) such as Google's Bard and Anthropic's Claude, new estimates of the impact that these technologies could have on the global economy have emerged. While, as mentioned in Introduction, earlier predictions by PricewaterhouseCoopers (PwC) pegged AI's contribution to the global economy at $15.7 trillion by 2030, post-ChatGPT estimates now lie at $25 trillion.[1] By comparison, this figure is roughly equal to the GDP of the world's largest economy: the United States. This is also a conservative estimate. The capabilities of LLMs are such that LLMs can now process huge volumes of input text or data at increasingly quicker rates – a process equivalent to digesting the contents of this book in mere seconds. With this, new heights of economic productivity are within reach, particularly for certain sectors and particularly for knowledge economies. While many fear job losses as a result of AI (a theme we will explore in Chapter 4), the capabilities of traditional white-collar labour are not going to be replaced by AI, they are going to be extended. These workers will have at their fingertips the potential to carry out their tasks at pace and at scale, which will generate a significant productivity jump.

As AI breaks through the new ceilings of wealth, what consequences will this have on poverty levels globally? It is mind-blowing to think that, with these new heights of productivity and economic gain from AI, we are sitting on the means to eradicate world poverty. The eradication of global poverty is estimated to cost $175 billion per year for 25 years.[2] With the $25 trillion that AI can help generate, this could solve world poverty five times over.

But, just like the PwC estimates, the new quoted figures for generative AI's contribution to the global economy offer limited details about how these numbers translate into economic gains for different regions of the world. The report from McKinsey in which these estimates are found indicates that the sectors likely to see the most economic growth from AI are the high-tech industry (e.g. Big Tech, space exploration, defence), banking, and retail. By contrast, the industry likely to see the least growth is agriculture, which is Africa's largest sector by a long way and the major source of livelihoods for the vast majority of Africans. Without a high level of saturation across all industries, AI's economic benefit will be narrowly confined and will have little direct impact on reducing poverty levels.

As the tech's billionaires get richer from the exploitation of vulnerable populations throughout the world, we are witnessing the largest transfer of income to the very top brackets of society. Globally, two thirds of all the wealth generated in the past four years has been amassed by the richest 1 per cent,[3] and the richest of them all are the new class of tech billionaires, equipped with the power, money, and influence to craft the worlds they want to live in.

In conversations in Davos or in moments when breakthroughs give rise to new heights of euphoria, AI may be spoken of as a gift that all humanity can enjoy and benefit from, as a solution to the greatest challenges that face the world. But, as we will find through closer examination in this chapter, the gap in AI use and development between rich and poor countries is deep and wide – and it is still deepening and widening. AI cannot just be developed by anyone anywhere. It requires a suite of resources, skills, and infrastructure that are simply not available to the majority of the world. China's desperation to get on top of the development of semiconductors, as reviewed in the previous chapter, is testament to just how complex and expensive it is to make this technology. Because of this, because of the exclusion inherent in the production of AI and in the system of returns, which ensures that only a select few benefit in stratospheric proportions, AI does not at present provide an opportunity to thrive that is available to all humanity. We are radically and wholly missing the chance to channel the benefits of this technology to the people and places that need it most. The technology is poised instead to drive further inequality between those who can develop and use it to their benefit and those who can't – or, worse, who are sacrificed in the value and supply chains of its planetary industry.

Historically, new technologies bring about economic growth and development. Many technologies are designed to do just this by boosting productivity: the sewing machine or the tractor, for example. Since the turn of the century, digital technologies have been a particularly powerful economic force globally. In the United States, the Internet is said to have supported the country's GDP to increase by 22 per cent a year since 2016. The US digital economy is now worth well over $4 trillion.[4]

But in Africa the economic gains from digitalization and increased Internet access over the past 20 years are patchy at best and, at worst, unequal. Perhaps the most significant example of a digital innovation that has meaningfully supported economic growth in Africa is Kenya's M-Pesa. M-Pesa is a mobile money platform that is now widely used across the country and that allows individuals and small businesses to safely and easily transact. It is reported to have contributed a 2 per cent growth to Kenya's GDP since its inception in 2007.

However, the success of M-Pesa is still a standout case. Although Internet penetration has increased dramatically across the continent, namely from close to 0 per cent in 2000 to around 45 per cent today, it is still way below the global average of 70 per cent. Moreover, access itself is not clear-cut, as research demonstrates a growing digital divide around gender: in Africa men have an increasingly higher rate of access to digital technologies than women. Rather starkly, statistics estimate that African women are 37 per cent less likely to have access to a smartphone than their male counterparts.[5] And, as in many aspects of life, the poor pay more than the rich for access to the Internet. In France you may pay $0.23 for one GB of data; in Namibia, you will pay at least $10. Currently, research into Internet usage on the African continent indicates that, while around 45 per cent of its population has access to the Internet, this access is only constant and reliable for around 20 per cent of Africans.

Added to this, Internet access in Africa is structurally precarious, relying on a network of undersea Internet cables, the establishment of which dates back to European colonialism. Not infrequently, Internet access is interrupted across much of the continent when there is damage to one of the deep-sea cables carrying Africa's connectivity. I've been in countless online meetings where colleagues across the continent could not connect for this very reason. One of the major undersea routes is the Equiano cable, owned by Google, which follows the notorious seafaring path of colonial discovery from Portugal to South Africa, stopping at the major ports along the coast of West Africa. It takes its name from Olaudah Equiano. Nigerian-born, Equiano was sold into slavery and shipped to the Caribbean, before making his way to Britain and becoming an important figure in the abolition of slavery through his writings. Esther Mwema, digital activist and fiction writer, is leading a creative campaign to rename Google's Equiano cable, which now stands as a bitterly ironic reminder of the all-too-present history of European colonialism in today's digital empire.

In this chapter we will explore the economics of AI and its relationship to the old European empires. We will examine why the costs and resources necessary to build AI models like ChatGPT are only available to the elite, and how the market dominance of Big Tech is driving a greater wedge between those places that use AI and those places that are used by AI. In order to better understand how these conditions of economic disparity have arisen, we will review the history of colonial economics and the creation of wealth inequality, looking in particular at how the West developed its technological capabilities through colonial exploits.

Billion-Dollar Enterprises

To understand the chasm that exists between the AI rich and the AI poor, it is useful to think about the resources needed to build today's most compelling AI system – ChatGPT. ChatGPT took off with a storm. In the first five days of its launch, 1 million people used it. Two months later, it had been used 100 billion times. It represents a major change in AI technologies, moving from computer intelligence capable of detection and prediction to intelligence able to produce and create human-like digital material. ChatGPT can manage your inbox not just by sorting emails by titles and keywords, but by generating a series of tailored responses to every communication. It can manage your investments and stock profile – and successfully so. Or it could write this book, if given some suggestions as to the kind of tone I want to convey and the type of research material I want to surface and analyse. This new class of AI technology was named 'generative AI'. And in one fell swoop ChatGPT demonstrated that a significant portion of human cognitive labour could be outsourced to a machine.

Generative AI has induced new concerns about the risks of AI. As described in the previous chapter, these technologies are the closest we have come to a general-purpose AI and AGI. They are much harder to regulate, as their potential uses are limitless. They have also raised profound questions about the nature of truth and the meaning of human culture, given that a machine is now able to write music and poetry or generate plausible news stories, videos, or images. The production of culture was a singularly human activity that, together with language, has historically been the crucial marker between the human and the nonhuman. That we are capable of producing artefacts and cultural materials that capture, defend, and transform the meanings, truths, and values we share between us has been a central part of human social activity among cultures everywhere. If AI represents a new species in the world, one capable of cultural production and truth-telling, what does it mean for being human today? While these questions are important to ask, it is also important to affirm that AI is a human-born technology that is remarkably adept at mimicry. In relation to art, it is, at best, a plagiarizer of human culture and truth.

But the creation of generative AI and of an AI system like ChatGPT is equally astonishing if we pause to reckon with the enormity of the resources, skills, and equipment it took to build it. ChatGPT was built by a company called OpenAI. OpenAI was founded in 2015, with the aim of creating an AGI that would benefit humanity. Its founding charter espouses the principle that its 'primary fiduciary duty is to humanity', namely to 'ensure' that AGI 'is used for the benefit of all, and to avoid enabling uses of AI or AGI that harm humanity or unduly concentrate power'.[6] While the group represented at the time a breakaway from the concentration of AI skills and power in Big Tech, it was shrouded in a gilded cloak of elitism: its humble beginnings had been underwritten by a $1 billion cheque from Silicon Valley heavyweights. Indeed, OpenAI has always made it clear that it would have to marshal substantial resources if it was to fulfil its mission – the admirable venture of building AGI for humanity. In March 2023, after the release of ChatGPT, which propelled OpenAI to household name status, Microsoft invested $10 billion in the company, incorporating ChatGPT technology into its web search engine, Bing. OpenAI is no longer a not-for-profit alternative AI company; it is a rich and powerful mainstream business seeking global market dominance, just like its competitors.

This kind of advanced AI could not have been developed anywhere outside the West or the advanced economies of the Far East. The audacity of the sheer scale of what generative AI encompasses cannot be overstated. Besides funding and commercial value equivalent to the GDP of some of the world's poorest countries and the acquisition of tech talent that reportedly costs as much as professional sports players, generative AI, just like ChatGPT, is conceived of by a daring imagination, which regards the entire archives of the Internet as a reserve to be raided. The minutiae of the material and data on the Internet come together into one great throbbing reservoir of information, which can be endlessly organized and given new permutations; and some of these may be trivial regurgitations, others may be fake new stories peddled around the world, while others still may be important discoveries of new therapeutics.

But to do this work of accessing and manipulating the Internet's global information requires power and resources beyond measure. To create an AI system like ChatGPT demands a range of compute resources that can process trillions of language-based data points from the Internet, then train the artificial neural networks with the help of these data points. Such resources include the semiconductor computer chips needed in their thousands for the general processing units (GPUs) and the nucleus of AI-driven computing systems, which process thousands of small tasks at once. They also include the software and the computer code, or language – such as Python, which instructs the processing undertaken by GPUs and manages the vast repositories of data. And, finally, they include the hardware infrastructure in which these activities are powered and contained, from the data centres with thousands of high-powered server units and the equipment needed to cool them to the undersea cables that transfer data around the world. And powering the development of ChatGPT requires a supercomputer.

Supercomputers differ from our everyday laptops and desktops in their degree of processing power. The units of computer processing power are called FLOPS – floating-point operations per second; and they measure the rate at which a computer can undertake one unit of activity, which might be an addition, a subtraction, or a multiplication of the floating-point numbers. The laptop on which I am writing this book has the power to process around 100 gigaFLOPS, which amounts to the power to process 10 billion units. Around the time ChatGPT was released, a supercomputer was a million times more powerful than an ordinary laptop: it could perform up to 100 petaFLOPS, in other words 100 quadrillion FLOPS. Microsoft invested $1 billion in its first supercomputer for OpenAI to train its AI models on, and this ultimately resulted in breakthroughs like ChatGPT. LLMs like ChatGPT reportedly require around 100 times more computer processing power; this is due to the huge number of data they are trained on and the complexity of their neural networks. In computing power alone, ChatGPT would have cost around $100 million to train.

This kind of computer is currently available only in a handful of countries around the world, the United States and Canada among them. Not even the United Kingdom has supercomputers on this scale – although, as mentioned in the previous chapter, Rishi Sunak announced plans to spend an eye-watering £900 million to build a British supercomputer. According to Vili Lehdonvirta, who teaches at Oxford University and has written on the material infrastructure of cloud computing, the United Kingdom is behind in the AI race and, without the requisite computing infrastructure, will not be able to catch up. Deftly named to suggest a weightless and immaterial expanse, the cloud is not the only place where data and software are housed; it refers instead to Internet-run servers, software, and databases enclosed in data centres that are often located in remote rural settings. Lehdonvirta argues that the United Kingdom may be without the necessary control over compute resources to ensure that cutting edge progress in AI development aligns with the country's technological objectives and values. This is because, without their own advanced computer infrastructure, UK AI firms will need to rely on renting compute power through the cloud computing services offered by the major tech companies. Currently Amazon Web Services, Microsoft Azure, and Google Cloud together make up 65 per cent of the global market share of cloud computing, maintaining a lucrative market position with which it is almost impossible to compete. As more and more pressure is put on industries and governments to adopt AI, and as it becomes harder and harder not to participate in the global ecosystems of data and in the Internet, in which AI is now inextricably implicated, the monopolistic power of cloud computing companies to engage in rent-seeking activities with just about everyone simply snowballs.

If the United Kingdom is not able to self-determine its direction and position in the global AI system, how will far less privileged countries fare? A review of global compute capacity, specifically the kind needed to develop advanced AI systems, has estimated that a third of this capacity lies in China, which is followed closely by the United States, with a quarter. The semiconductor embargo is set to ensure, however, that the

United States and its allies get ahead in the supercomputer race, leaving China unable to increase its national compute power.

Between them, the regions of the global South hold less than 3 per cent of global compute capacity. Africa – home to almost 1.5 billion people – has only 0.2 per cent. At present, with the kind of computing capacity available in Africa, South America, or Oceania, it would take hundreds of years to catch up with the advances that have been made with LLMs in the West and developed East. The UK government's recent commitment to invest in securing a national exascale computer is indicative of the kind of funding required to meet the levels of compute capacity one needs in order to compete in the global AI race. As I have already pointed out, such a sum is equivalent to the national GDP of a number of the world's poorest countries. This means that the vast majority of countries in Africa, South East Asia, and Latin America and in many small island states are almost completely excluded from ever being able to catch up and take advantage of this technology on their own terms. The only path left is to rent computing services, and data, from the cloud computing giants.

But the market dominance that these companies boast is not simply an advantage in terms of resources, whereby others could catch up if they were able to mobilize the resources to build comparable compute services. These companies, like many digital platform providers before them, have lucked out on first-mover advantage. Expressing a new model of global business, first-mover advantage describes the critical lead that first players in a new market, typically digital, can obtain. By offering a new service that works at scale and through a network, for instance Uber or AirBnB, these companies can effectively create new markets and almost entirely dominate them. It is then almost impossible for new market entrants to offer competitive services; without an extensive network of linked cars or rentals, a competitor to Uber or AirBnB would not be able to offer anything like the breadth of options and availability that customers have come to enjoy. Apple has done this very well with its suite of linked devices. And software packages such as Microsoft 360 and the Google suite have established markets locked in to their systems and services. These market tactics rely on and embed a logic of empire, as one massive company dominates over billions of users, establishing soft dependencies that seem almost impossible to escape. Anyone who has tried to delete their Facebook or Gmail account can tell you this.

A few years ago, before the rise of cloud computing, no one deployed first-mover advantage better than Facebook. Facebook was the first major social networking platform, amassing more than 3 billion users from around the world in less than 15 years since launch; its business model involves monetizing your attention span and data. By connecting vast networks of people and making them share the details of their daily lives with one another, Facebook monetizes these people's attention and data; and it does so through algorithmically selected personalized adverts. The bigger the company's network, the more precise the data profiles it can built and tweak; the more personal the information shared, the more the site can act as a global pulse check, assessing how a (or any) given group might feel about a current issue, be that a new product on the market or a presidential candidate. This has, of course, led to some significant instances of harm. Facebook was involved in the Cambridge Analytica scandal, which reportedly influenced the outcomes of both the US presidential election of 2016 and the UK Brexit referendum; and it also had a hand in the 2016 genocide of Rohingya Muslims in Myanmar, where its algorithms promoted content that incited violence against the group. In the same year these events took place, the *Economist* published an article on the company and its founder, Mark Zuckerberg, discussing its 'imperial ambitions' and how AI would augment Facebook's hold over the world. An image of a statue of Zuckerberg as Roman emperor (*Marcus Zuckerbergus*) accompanied the piece, with the motto *Coniunge et impera* ('Connect and rule', the reverse of the famous *Divide et impera*) engraved on the plinth.[7]

In 2021, as the world's dependence on digital connection had so desperately deepened during the Covid-19 pandemic, Zuckerberg announced a rebranding. Facebook would become Meta, and with it a new digital realm – the Metaverse – would be launched. The Metaverse was to be a virtual online world where people could interact and do business. The move represented a second phase of Zuckerberg's imperialistic ambitions. With much of the world already conquered by Facebook's lure, the time had come to create new territories into which to expand and rent out space. Billions of dollars have been spent by companies, from Adidas to Gucci, on virtual real estate, but Meta itself has lost out: no one stays for long in the Metaverse.

While Big Tech's industry is global – Africa, Asia, and Latin America constitute the largest regional user blocs of Facebook, for example – there is little direct economic benefit to be gained in countries where business is entirely digital and has limited footprint on the ground. This is because these digital industries do not depend on local skills that generate employment; nor do they depend on local property and real estate. In addition, these global digital giants constitute a major barrier to new market entrants, and especially local businesses. For new market entrants it's getting harder to counteract the first-mover advantage of large imperialistic tech platforms whose market share is both increasing and becoming entrenched.

Investment patterns, too, are fuelling market dominance. In the United States, recent reports demonstrate that, while AI research and development are increasing, there are fewer investments in AI companies.[8] This reveals a tech market that is narrowing rather than opening, which means that the economic benefits accrue to a smaller few and the barriers to entry for new players are set too high for AI companies without serious resources and capital to compete.

Even without compute power, the cost of the data needed to train AI raises another barrier to entry for under-resourced AI firms. Where each piece of data costs, and there is limited access to comprehensive public datasets, the efficacy and accuracy of the AI system that is being built will be compromised, so the system will be ineffective and harmful.[9] The other data issue that has material consequences is that, as AI increases its demand for more data to refine its models, spaces must be found to store the vast seas of data, which grow by the millisecond, and the extensive server systems required to power them. As the material volume of data in the world increases, new places are being sought to house them and the server farms. Underwater data storage facilities are being established off the coast of Scotland, for example, where the cold Atlantic waters can operate as an effective cooling device for the excess heat produced from the computational humdrum inside. And a company named Cloud Constellation Corporation is developing satellites marketed as offering up to five petabytes of space-based secure data storage and services. This company boasts that it will use 'laser communication links between the satellites to transmit data between different locations on Earth'.[10] Indeed, one should not understate the significance of the sheer quantity of data that are now produced in the world and that require an end point in the network to be physically located.

It won't be long before Big Tech turns to the cheaper land spaces across the majority world, to house there its swelling heaps of data. This will not be without cost to local communities and environments. Data centres require mega volumes of water and energy to be cooled and powered. Here is how Rachel Coldicutt, founder of the people-first technology, research, and advocacy group Doteveryone, explains to parliament how communities in England are affected by Big Tech's data needs:

LLMs are particularly notable for their excessive water consumption, as a result of liquid cooling used in data centres
– with an estimated 700,000 litres of water required to train
OpenAI's GPT-3 in Microsoft's data centres. This will contribute to worsening droughts amid wider climate collapse. The
UK's water infrastructure is ill-equipped to support the growth
of new data centres, due in part to the lack of new reservoirs.
As such, data centres are already competing with people and
agriculture for scarce water resources: in three boroughs in
west London the grid has run out of power to support new
house building due to the number of data centres being built
along the M4 corridor while in Cambridge, water scarcity is
one factor slowing down the house-building ambitions that are
essential to turning the city into a 'science capital'.[11]

The situation is made worse by a set of security protocols
that render data storage inherently inefficient and unsustainable. Such protocols require that data be stored in multiple
sites in order to ensure security and availability even in the
event that one dataset is destroyed. If you have an iPhone, your
personal photos will be stored on your phone, on a data server
on a data farm somewhere in the world, and also in another
backup data centre, somewhere else in the world – all to
ensure you never lose them. Some of the major breakthroughs
in the history of computing have happened in the course of
developing smaller and more efficient ways to store data and
information, the cloud being the most significant innovation
in recent years and nanotechnology being on the horizon. But
these breakthroughs have been motivated by a desire to do
more with more rather than to make the industry sustainable.

If Big Tech starts exploring cheaper land in places across the
majority world in order to house its data farms, this will place
even greater demand on sources such as water and energy, which
are already scarce. The energy deficit across Africa is impacting
people's access to advanced technologies. A colleague of mine

had conducted a digital capacity-building program in rural Cameroon that involved offering smartphones at a fraction of their retail price. The smartphones, however, held little appeal to the communities to which they were offered. With a battery life that lasted just one day and a device that would need to be used by multiple people, the smartphone was simply not practical by comparison with a standard feature phone, whose battery could last a week. Here, as in many places across the continent, reliable access to electricity is a luxury that often can be attained only through off-the-grid solutions. Recent statistics from AfroBarometer indicate that not even a half of Africa's population has reliable access to electricity.[12] Without electricity, much of the potential benefits of AI are wildly out of reach.

This reality was reinforced to me when I taught a course on digital humanities to master's students in Yaoundé, Cameroon's capital city. Our focus was on exploring how digital tools could provide new means of developing and understanding local cultural expressions and histories. All the students were digital entrepreneurs of one kind or another, involved in building educational resources for children in local languages, or video games that reflected cultural norms and idioms. We often came back to the defining question of digital access – access not just to the Internet but, critically, to reliable electricity services. Some students would tell stories of how they were able to speak with their parents, who lived in rural areas of the country, only every six months or so. This was because the feature phones their parents had access to would be charged only once or twice a year, when the parents had access to electricity. These were not simply limitations within the electricity infrastructure; they were caused by the fact that the national grid regularly ran completely out of power for rural areas. Although in South Africa, where we live, we frequently experience loadshedding – the electricity is periodically turned off in one area for a set period, often up to 12 hours a day, in order to reduce the load on the national grid – those students' stories were a stark reminder of how privileged we were in South Africa to have fairly reliable electricity access.

Energy is of course the basic unit of industry. Without steady Internet access and reliable electricity, let alone compute power, the saturation of AI across industries, enough to markedly increase the total factor productivity and GDP, is simply not possible. The energy crisis across Africa is pervasive and potentially worsening, as the existing infrastructure and capacities cannot meet the rising energy demands. Libyans, for instance, regularly face 24-hour blackouts, and there is a number of contributing factors such as internal and geopolitical conflict, corruption, and infrastructure theft. In South Africa, electricity theft is a major issue that the national energy provider is struggling to manage. This is primarily a symptom of poverty and takes either the form of cable theft, where electricity cables are earthed and dismantled, then the valuable copper and aluminium used to encase the wires are extracted and sold for a measly profit; or the form of illegal connections, where electricity cables are rerouted to places in need.

Digitalization and AI place enormous energy demands on these already fragile systems, and will do so even more in the future. Data centres located in energy-poor contexts will require as much energy as an African metropole, potentially causing a whole range of disruptions, if national energy systems and infrastructure are not updated and better capacitated.

Colonial Economics

That Africa and other countries of the global majority are without the resources they need to advance local AI ecosystems is not without reason. Colonialism left the colonies it claimed ravished and poor. Primarily, colonialism was big business: a global enterprise of economic extraction. European imperialist agendas sought to enlarge the opportunities for national economic growth through the appropriation of land, labour, and natural resources from the colonies, establishing new trade routes and markets with which to sell back much of what they took at extortionate prices.

The regression of India's economic wealth before and after British rule tells a stark tale of economic exploitation. India was colonized by the East India Company, which was to play a major role in shaping the birth of global capitalism during the colonial era. Dadabhai Naoroji, three times leader of the Indian National Congress under British rule, estimated that, from 1833 to 1872 alone, Britain amassed a wealth of over £500 million from its exploits in India.[13] But India itself was losing out. The statistics demonstrate that under British rule, in the first half of the twentieth century, India's GDP was stagnant or even regressed. Mortality rates were high, literacy was low (and, for women, effectively nonexistent), and the colonized population was organized into productivity blocks that drove economic accumulation for Britain alone. Indeed, while India and the other colonies were being depleted, their imperial masters amassed such a significant portion of wealth from foreign-owned assets that they were still able to profit while maintaining a trade deficit. Thomas Piketty explains:

> These very large net positions in foreign assets allowed Britain and France [the world's foremost empires at the time] to run structural trade deficits in the late nineteenth and early twentieth century. Between 1880 and 1914, both countries received significantly more in goods and services from the rest of the world than they exported themselves. ... This posed no problem, because their income from foreign assets totalled more than 5 per cent of national income. Their balance of payments was thus strongly positive, which enabled them to increase their holdings of foreign assets year after year. In other words, the rest

of the world worked to increase consumption by the colonial powers and at the same time became more and more indebted to those same powers. . . . The advantage of owning things is that one can continue to consume and accumulate without having to work, or at any rate continue to consume and accumulate more than one could produce on one's own. The same was true on an international scale in the age of colonialism.[14]

In effect, a global shift took place in economics: from capital earned through labour and productivity to wealth amassed through assets and rent. This created the conditions for global wealth inequality, where those who own assets can profit from renting them and accumulating wealth, without losing the ability to earn an income through labour, or without needing to altogether. For those without land or property to rent, the only source of income was the labour they could offer, or else they would enter into punitive debt agreements. Through this model, colonialism not only constituted the most severe historical example of systemic profiteering in the modern world but also laid the conditions for the continued extortion of colonized land, populations, and groups long after its own end.

But in the process of formal decolonization, when administrative power was handed back to the colonies, colonial powers emptied out the last colonial coffers and established new forms and practices through which they could continue to extract wealth from the colonies. South African scholar Mpofu-Walsh writes, for instance, of how apartheid did not end with the dawn of democracy and the election of an African president in 1994; apartheid was instead privatized.[15] In the years preceding the handover, the enormous wealth of the white minority apartheid government was carefully transferred into the hands of industry giants, from mining to finance, who had long begun to anticipate the moment of majority rule. When the new constitution of the Republic of South Africa was passed in 1996, after years of deliberation across both sides of power, the old

and the new, a right to private property and to the expropriation of land *with* compensation was enshrined, ensuring the
centralization of national wealth in the hands of the minority
white population. In years to come, as opposition voices rose
to contest the failure of the new democratic dispensation to
enact a redistribution of wealth in the country, a slogan came
to dominate the public discourse: it was calling for the fall of
white monopoly capital.

In the former colonies of France, a formidable system of
economic control was enacted in 1945 to guarantee continued profit from the colonies back to Paris: its instrument was
the CFA franc. In their book *Africa's Last Colonial Currency*,
Fanny Pigeaud and Ndongo Samba Sylla described the sovereign power that France continues to wield over a currency
used by 14 French-speaking African countries.[16] Although the
currency was given an Africanized face through the African
leaders who adorn the various dominations of notes, these
notes are printed in France and their exchange or sale must
be authorized by the French Treasury. For Pigeaud and Sylla,
the CFA franc gives France the mechanism through which to
'manage its economic, monetary, financial and political relations with some of its former colonies according to a logic
functional to its interests'. It is, they write, 'the most powerful
weapon of the "Françafrique", this peculiar neocolonial system
of domination that the French state established on the eve
of the independence of the former colonies, with the precise
aim of preserving the advantages of the colonial pact'. These
enduring systems of control disallow meaningful economic
sovereignty for African nations. They critically undermine
the nations' capacity for economic development on their own
terms. The downstream effect is the continued impoverish-
ment of African people and communities.

When the colonizers expropriated the wealth of the colonized lands, the West equipped itself with enough resources
to build new industries and technical capacities. The West's

rise to the apex of technological and AI capabilities depended heavily on the colonial order of things, although this path was slow and protracted. It was slow not just because of the years in which the US Defense Advanced Research Projects Agency (DARPA) and other state agencies funded early AI work, but also because of the slow centuries of scientific endeavour and its application to industry and modes of production, as well as because of the slow development of the correlative human skills needed to work these industries and its new machines. As the Industrial Revolution took hold in the West in the nineteenth century and scientific discoveries were applied to develop new technologies and advance the range and capabilities of industry, raw materials and manpower had to be sourced to build and fuel this growth. The new discoveries of land in Asia, Africa, and the Americas offered vital access precisely to the assets that were needed. An important book for understanding the historical dynamic of the rise of technological industries in the West comes from the Guyanese writer and activist Walter Rodney, who produced it during his professorship at the University of Dar-es-Salaam in Tanzania in 1972.[17] What Rodney explores is how colonialism was a key driving force behind the development of technology and related expertise in the imperial centres and provided the labour and raw materials with which new industries could be built, while it kept production costs low through exploitation.

But all this came at the cost of a regression in Africa's own technologies. Indeed, Rodney explains how slavery and colonialism robbed Africa of the opportunity to make technological innovations and develop skills alongside its productive capacities, abruptly halting indigenous technological forms and practices. The degeneration of iron smelting skills and traditions was perhaps the most significant technological loss, taking with it the ability of African communities to fashion and produce their own tools, weaponry, and cultural materials.

Besides the loss of technical traditions, the Atlantic slave trade took much of Africa's youth, skilled men, and labour force as slaves. The trade in goods that continued between the West and Africa was in raw materials, ivory, and other commodities that did not require technical skills or productive capacities and so were not locally cultivated. These practices of extraction are dominant in global trade and supply chains that take raw materials from the continent (and also from other places from across the majority world) and refine them for industry in factories in the West, where the technologies and skills have been developed. This system denies African communities and entrepreneurs the ability to build their own productive industries and to beneficiate – that is, add value to – the raw materials extracted from their lands; these are cast instead in the eternal role of the miner. As Rodney explains, 'it was the advance of scientific technique in the metropoles' that, more than anything else, 'was the cause of the great gulf between African and Western European levels of productivity by the end of the colonial period'.[18] Today, global supply chains – including the supply chains that create the hardware upon which AI is ultimately built[19] – are still structured along these global lines of division, delivering western states, and now China too, the true wealth of postcolonial lands and creating a crisp global division of technical skills.

AI and, later, its mega industries of scale and wealth arose against the historical and global division between scientific and technological capabilities. These industries are built on centuries of technical infrastructure development, skills acquisition in computing and data science, and sociocultural buy-in of the Enlightenment doctrine of technological determinism – that is, the idea that increasing technological development is the natural path to progress for advanced civilizations. The sheer size and complexity of what is needed to build an AI system is almost unfathomable in comparison to the resources available in most postcolonial nations, let alone the rate at which it needs to be adopted if we are to see the kind of economic growth that is being projected. This is no accident. It is the product of centuries of scientific and technical progress that deliberately left the colonies behind, while exploiting their resources to power this very progress.

Technology Diffusion

Supposedly the economic benefits of AI will reach everyone everywhere. When richer nations claim to be leading the world in AI development, the assumption seems to be that everyone else is eventually brought along with them. This thinking is built on ideas around the power of technology diffusion: over time, technology becomes cheaper and easier to use, translating into greater access for more people.

I heard this narrative sung recently by Mustafa Suleyman, co-founder of Google DeepMind and recently appointed CEO of Microsoft AI, who spoke about how universal access to AI was no longer a concern. Instead, he stated, the world is witnessing a rapid diffusion of AI everywhere, and almost everyone has access to its benefits. Our problem, he said, was to control access, not widen it. In my experience, this is not the case. Where smartphone penetration is low, access to AI-driven apps or resources that might assist individuals is impossible – as we saw in Cameroon. Those without direct access may interact with AI systems through their government's use of the technology. Across much of Africa and the majority world, governments are turning to AI to drive efficiency as part of the rollout and monitoring of government services. But there is little evidence to demonstrate that this brings much benefit to citizens. There is instead some evidence of harm, which we will explore in Chapters 5 and 6.

Even if compute processing becomes cheaper and AI gets less expensive to build, its diffusion across industries and its contribution to increasing a nation's total factor productivity (a traditional economic indicator of national technological progress and its relationship to economic productivity) will stall without the incentives, skills, and capacity to adopt it.

Successful technological diffusion depends on a number of factors. Certainly it depends on technology becoming cheaper and therefore lowering cost barriers to access, and on technology becoming easier to use, so that more people may be able to employ it. But it also depends on what is needed – what is called 'absorptive capacity'. Many of the infrastructural issues (electricity, Internet, compute power, digital infrastructure) described in this chapter – issues that represent barriers to AI development and adoption outside the West and the richer countries of the East – point to a limited absorptive capacity for AI, in Africa in particular.

This problem is amplified by lower levels of education and skills, technology diffusion being most effective where education and skills levels are high. Indeed, the rapid rate of adoption of AI in high-income countries is due to the competitive need to reduce labour costs, where automation offers cheaper labour (more on this in Chapter 4), and to boost the productivity of already highly skilled workers through advanced technologies. In Africa, where the labour force is less skilled, industries do not have the same incentives to employ automation to reduce labour costs that are already low. Additionally, a lower-skilled workforce is far less able to adapt to the use of new technologies in daily tasks – at least until AI technologies are significantly easier to adopt in relation to current worker activities and tasks. As we saw in the opening of this chapter, the latest generation of AI technologies – generative AI – is not predicted to offer significant economic benefits to Africa's largest industry, agriculture. This, in fact, points less to what Africa lacks in the global landscape of resources and skills and more to the fact that AI has been designed and developed to serve a suite of productivity needs of well-developed knowledge economies. For large parts of Africa, AI may simply not be fit for purpose.[20]

Part of the reason why M-Pesa has taken off as it has in Kenya is that the technology is fit for purpose. It was designed to meet people at the stage they were at, providing safe and reliable ways to transfer money and transact for anyone with a phone (and it doesn't need to be a smartphone). M-Pesa is a good example of how digital technologies need to be designed for the context in which they will be used in order to support meaningful technological diffusion. More efforts are needed across the African continent – and across the majority world – to explore which forms and types of AI technologies are most fitting alongside current social and economic activity and to design them accordingly. For example, in his engagements with me, Moustafa Cissé, who headed up the Africa Google AI Lab in Accra, Ghana and founded the first master's course on AI at an African institution, has spoken about the value of developing voice-led AI in local languages for the continent. Noting varying levels of literacy, including technical literacy, voice-led AI is likely to be the easiest form of AI for most people to interact with. Think Siri, or sending voice notes to an AI application that could respond in your local language with advice on when to harvest or where to sell a crop yield. In explaining how this could be developed practically, Cisse pointed to the extensive archive of radio data in local languages available across Africa, and explained how this could be digitalized and used to train AI models in these languages. AI may very well have a place here, but let's not assume that its successful use will look the same as in other parts of the world.

Technology diffusion has not always been so difficult. In previous technological revolutions, the diffusion of technology has taken place through the establishment of infrastructural networks that, arguably, can be used by anyone. Examples include railways, the Suez Canal, the telegraph, or even global Internet networks. But technology diffusion for AI rests on a more complex set of preconditions that are concentrated in the hands of a few. What all this means is that richer nations are reaping and will continue to reap the economic benefits of a technology that has, by and large, been developed to do just this, and the $25 trillion set to accumulate to the global economy will oscillate within these spaces alone, widening the global inequality gap.

It would be a mistake to assume that the benefits – and, as the focus is here, particularly the economic benefits – of AI trickle down to everyone. Hence as much energy as is being spent on trying to build AI capabilities and capacities across Africa should be directed to redistributing the wealth of AI to address global poverty.

3

The Material World of AI

Accra – Ghana's major metropolis – opens out along the Gulf of Guinea on the west coast of Africa. The low-lying city takes its name from the word for 'ant' in the local Akan and Ga languages, referencing the unmissable red anthills that tower over the hinterland of the surrounding Accra Plains. These anthills are culturally significant *aklabatsas* – sacred groves or enclosures wherein the spiritual and the human worlds meet,[1] symbols of human cooperation, communality, and organization.

Accra has an embattled history of colonial conquest. In the early sixteenth century, Portuguese invaders occupied a site in this region before their stronghold was seized in 1578 by the Ga people, who came down inland from Togo and Benin, eastern regions of present-day eastern Ghana. By 1642, the Dutch had established a trading post on the site and expelled the Portuguese. This was before British occupation, which came in the second half of the seventeenth century. In 1877 the British moved their colonial capital from Cape Coast – the chief slave port of the Gold Coast, jutting out of the Ghanaian coastline – to Accra. Eighty years later, Ghana became the first African nation to win independence from its colonizers. This came about under the leadership of Kwame Nkrumah, Ghana's first post-independence prime minister and president. In 1958, a year after independence, Accra was to host delegates from liberatory movements across the continent to the first All African People Conference, where the future of Pan-Africanism and the continent's decolonization was fortified. The slogan for the conference was 'Hands off Africa'.

In the heart of Accra, adjacently to Odaw River, which flows through the Korle Lagoon into the Atlantic Ocean, is an area called Agbogbloshie. It is home to one of the world's largest sites of e-waste. Every year, hundreds of thousands of tons of defunct electrical goods – smart devices and technologies made obsolete by the rapidly developing tech industry – arrive from Europe, the United States, Australia, and other wealthy nations at the port town of Tema, an hour outside of Accra. A UN report estimates that 50 million tons of e-waste (electronic and electric waste) is produced each year,[2] an amount that is steadily increasing as new smart devices enter the mass market and the technologies of yesterday are outdated by newer and faster models. Much of this is dumped in Agbogbloshie. The global production of e-waste mirrors the contours of economic privilege. The average amount of e-waste produced in the global North is around 20 kilograms per person each year, while on the African continent, where much of the e-waste ends up, the estimated figure is under 2 kilograms per person each year.[3]

More than 100,000 people live in the electric wasteland of Agbogbloshie. Many have migrated down from the north of the country, in search of work and opportunities in the burgeoning city. Those residing in Agbogbloshie seek out a living by sifting through scraps of electrical hardware and wires, plastic casings, and circuit boards; they look for traces of embedded metals such as copper and aluminium, which can be sold on. These people's work of dismantling the West's old electronics and reclaiming the metals that, as we will see in this chapter, have often been mined from the soil of their own continent is physically grueling and dangerous. As they rummage for scraps of value in the electrical rubbish, workers are exposed to countless harmful toxins. These toxins poison the Odaw River and the groundwater of Agbogbloshie; they carry carcinogens and other harmful substances to nearby communities and farmlands, and they affect a still greater number of people through contamination of crops and water supplies, while above their heads a heavy black cloud broods in the relentless heat produced by the burning of electrical cables that pollutes the skies of Accra. Studies examining the effects on the health of those who live and work in Agbogbloshie have found that people on the site experience damage to their eyes, skin, and hearing, as well as longer-term increased risk of liver and DNA damage and reduced heart rate and lung capacity.[4]

The living conditions of families that reside in Agbogbloshie are dire. In the slums that have been erected, chickens and other livestock pick about through broken smartphones and the carcases of old refrigerators. A study determined that the chicken eggs of Agbogbloshie had the highest incidence of hazardous brominated and chlorinated dioxins ever found in free range eggs.[5] These toxic eggs may be all that the families in the area have available to eat – a situation with critical consequences for pregnant women and for the health of children and Ghana's future generations. Given that the air pollution and groundwater contamination are also caused by the daily dumping and salvaging that take place in Agbogbloshie, the area constitutes a major health risk and environmental concern for the residents of Accra.

The goods that arrive in Agbogbloshie are often not declared as e-waste. Instead they are labelled as used electrical goods, although they are defective or, to be fixed, require resources that are not locally available. By not declaring the items as e-waste, those countries that export their outmoded electronics can circumvent the requirements of the Basel Convention on the Control of Transboundary Movements of Hazardous Wastes and Their Disposal, which has prescriptions for the safe dumping of e-waste – and Agbogbloshie would certainly not meet the requisite standards. Almost all countries are signatories to this instrument of international law, established in the early 1990s after a general outcry against the environmental wreckage caused by the dumping of hazardous waste in developing countries. The Basel Convention document requires countries that import or export hazardous waste to take steps to ensure that the movement and disposal of such waste has minimal negative impact on the health of local communities and their environments. However, it is estimated that about 30 per cent of the 1.3 million tons of e-waste carried out each year from Europe are illegal exports mis-labelled as used goods.[6]

Agbogbloshie exposes a part of the AI supply chain that is rarely mentioned. A graveyard of broken smartphones and computers, Agbogbloshie demands that we reckon with and take accountability for this sheer mass of material. To put it into perspective, the amount of electrical waste that arrives in Agbogbloshie each year is equivalent to six Titanic ships. And the figure is only set to increase, as new technologies are produced and new systems of connection require upgrading. In 2022, for instance, the European Union has passed a directive to standardize electronic charging devices such as phones, headphones, and tablets. When this comes into effect, millions of charging cables would immediately become defective. What happens to the old chargers? It is likely that they, too, would end up in places like Agbogbloshie, if the situation is not carefully managed otherwise. Certainly, as more and more 'things' get connected to the AI-driven Internet of Things (it is estimated that by 2025 this number will be about 22 billion), their older and less smart models will turn into garbage, to be dumped in parts of the world forgotten by those with smart cars and smart fridges.

The work taken to produce AI, including the waste it creates or to which it contributes, is carefully hidden by an industry that prefers to depict AI as a clever add-on to connected data systems already in existence. The 'stuff' and raw matter it is made from must, it seems, remain invisible: its materiality must be hidden in order for its rapturous promise of ease and efficiency to be preserved in the eyes of those who use AI and benefit from it, who do not want their enjoyment of AI to be tainted by the violence hidden in its production. This fallacy is bound up with an old colonial dream: the colonizer enjoys an easy life, remaining blind to the fact that his comfort is predicated on oppressing the colonized through appropriation of their labour and expropriation of their land and resources. His utmost privilege dulled his awareness of the cruelty felt by the colonized. Just like in the world of AI today, the fallacy had to be upheld at all costs, so that the dream of human progress and betterment can live on. Surely AI, like European colonialism, can only be making the world better.

In this chapter we will explore the disjuncture between how AI is imagined and the material reality of its production. We will examine mining activities that, from the Democratic Republic of Congo to the Atacama Desert in Chile, feed the AI value chain. In a carefully constructed global arrangement, which depends almost entirely on well-trodden colonial paths of power and exploitation, the unsightly materiality involved in the production of AI is outsourced to the global South, leaving untarnished the sleek image of a high-tech and comfortable life that is presented as the hallmark of modern western living. In short, we will learn about what AI takes and does not give back.

AI's Beginnings

We imagine AI in the form of intangible telepathic exchanges across a computerized network in the ether. Its concrete materiality is hard to conceive of beyond the cylindrical shell of an Alexa device, the muscular casing of C-3PO, or *The Terminator*. In reality, AI exists within a massive material ecosystem that stretches across the planet, placing ever greater demands on the earth and its people, with costly human and environmental effects. The layers of AI's material body run deep. AI feeds off mammoth quantities of data in order to train, retrain, perfect, and customize its algorithmic machine-learning models. These data are acquired through many sources such as CCTV footage, or are scraped off the keystrokes of a smartphone. But the Internet has revolutionized the capacity of AI by providing instantaneous access to increasingly deeper pools of data, creating the foundation for frontier AI technologies like ChatGPT. For this reason, the entire materiality of the Internet – the billions of devices and smart objects around the world that produce new data at every turn and click, the semiconductors that move data from devices and into cables, to be carried across our planet, the cables below the ground and sea, the server farms that power this complex ecosystem, and all the data centres in which the data are deposited – is, all of it, essential to the operations of AI.

The hardware used to create and expand the global digital configuration is fashioned from raw materials extracted from the earth's crust, refined and manufactured into the electronic items and cableways that eventually make their way to places like Agbogbloshie. These activities place ineffable strains on the planet, causing harm to local communities and the environments in which they live, and even inciting illegal and violent activities carried out to meet the ever-deepening demands of AI.

Increasing attention is being paid to the planetary demands and costs of AI. Kate Crawford and Vladen Joler, for example, offer a stark visual explanation of the human and material resources involved in the life of a single Amazon Echo – from the mines and the manufacturers to the Internet providers and the domestic infrastructure, then to the recovery and disposal of discarded devices – demonstrating the sheer volume and complexity of the supply chains involved in its production (see Figure 3).[7] But what we have yet to fully grasp is how the planetary apparatus of AI depends upon the global conditions that colonialism has created: surplus workforces, supplies of raw materials, and stretches of land located in former colonies – all considered ready for the taking. We will begin at the earth's crust.

As technology becomes more complex and connected to AI systems, a larger amount and a wider range of raw materials are required for building and maintaining technologies and the infrastructures on which they operate. With their light-emitting diode (LED) screens, rechargeable batteries, and wireless capabilities, smartphones for example require a far more complex array of elements than earlier landline telephones – and these include rare earth elements (REEs). Here are some of the most important elements used in smartphones and the systems and devices of the contemporary digital era: silicon – the most widely used raw material for semiconductors; cobalt and lithium, used in rechargeable batteries; germanium, used in infrared wireless technologies; indium, selenium, and gallium, used in LED screens; and what is known as 'the three Ts' – tantalum, tin, and tungsten, whose properties are essential to various electronic applications.

Much of the earth's supply of these elements is found in China, Brazil, India, Thailand, and especially in a number of African countries: Rwanda, Nigeria, Ethiopia, Burundi, and the Democratic Republic of Congo (DRC). The mining of materials such as tantalum and cobalt is a major source of conflict and indigence in mining regions across the world. Their increasing value and the demand for them were caused by the digital transformation of the world. The demand for cobalt, for example, has increased more than 90 per cent every year since around 2010. In Brazil there has been a radical increase

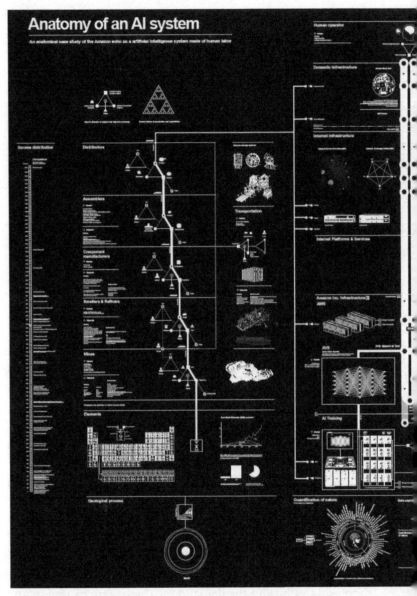

Figure 3   Crawford and Joler's 'Anatomy of an AI System' for an
Amazon Echo
Source: Crawford & Joler, https://anatomyof.ai

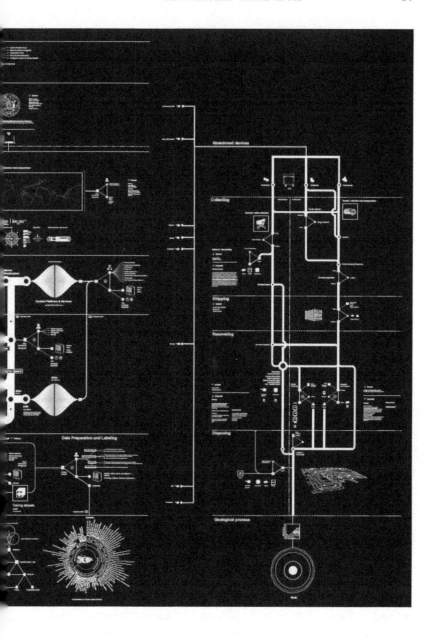

in illegal mining activities in the Terra Indígena – territories declared traditional indigenous land in Article 231 of the Brazilian Constitution and supposedly protected from illegal logging, mining, or settlement activities. Raw materials mined in these regions include manganese, used in rechargeable batteries, and gold, used as electroplated coating in computers and smartphones. Mining in these regions has untold effects on local indigenous communities; it disturbs the biodiversity, causes soil fragility, and affects subsistence farming. More immediately, mining activities have a deleterious effect on local teenage girls, who are seized from local communities and submitted to forced labour and marriages, as well as to sexual violence at the hands of mine workers and the makeshift communes they establish during months-long excursions into remote regions. In those far-flung places, the girls have no protection and no way out, even if formal complaints are registered with the Brazilian government and the ministry responsible for indigenous affairs.

In much the same way, mining in central Africa is today, and has been historically, a major cause of political instability and civil war, particularly in the Democratic Republic of Congo (DRC). For years, peace efforts in the DRC have been marred by the greedy attempts of foreign companies, African leaders, and local militia to control the wealth hidden within the country's soil. In the mining regions of the country, children are being recruited in the scramble for cobalt and REEs, their small bodies exploited as they are made to crawl into narrow mineshafts and pick through unearthed rubble. In 2019 a significant complaint of human rights abuse was lodged by an international human rights group at a federal court in Washington, DC against some of the major buyers of cobalt – Tesla, Amazon, Apple, and Microsoft among them. The complaint was made on behalf of the families of children who worked in cobalt mines in the DRC and had died or suffered lifelong injuries as a result of their perilous work. In 2021 the case was dismissed, as the judge determined that there was insufficient evidence to tie the incidents to the named companies. While this confirms how urgent it is to obtain evidence that can untangle the knotty links between the global supply chain of AI and advanced digital technologies and to bring such wrongs to light, it is also revealing of a system of justice that is skewed towards power and privilege. Just like the young girls from the mining regions of Brazil's Terra Indígena, these children cannot depend on the formal systems of justice to protect them.

In Africa, the mining of REEs and elements used for information communication technologies (ICTs) constitutes a far more significant percentage of the total national mine production than in China or in other parts of the world. In the Great Lakes region of Central Africa, where the DRC is located, tantalum production is the largest contributor to the national economy of Rwanda, while the economic value of these elements in the DRC is over $300 million. Yet much of the mining activity in this region and in other mining areas in the global South fails to yield meaningful benefits to local communities and economies. The DRC, for example, is the richest country in the world in terms of the raw minerals found in its soil, which are valued at $24 trillion (and more). It is staggering that the DRC is at the same time one of the world's poorest countries and three quarters of its population – around 60 million people – live in extreme poverty, on less than $2 a day. This disconnect is fundamentally linked to the colonial history of the country.

The Battle for the Heart of Darkness

The Great Lakes region of Africa is the part of the world that perhaps most tightly gripped the imperial imagination. European explorations in Africa were initially limited to the coastline

Figure 4   John Thomson's 1813 map of Africa, with central Africa
as 'unknown parts'

that spread from the Cape of Good Hope on the southern tip
of the continent to the coasts of West Africa. Much of West
Africa had been penetrated by Europeans as part of the transatlantic slave trade from the fifteenth century onward. But,
to Europe, central Africa remained an impenetrable enigma.
By the nineteenth century, tales of European explorers who
breached the interior of the continent had become legends;
David Livingstone and his search for the source of the Nile is a
prototype. Expeditions would set out from the port of Boma on
the western Atlantic Coast of Africa, travelling up the Congo
River into what came to be known as the heart of darkness,
after the publication of Joseph Conrad's eponymous novel at
the turn of the twentieth century (see Figure 4).

Many European explorers were prospectors for colonial enterprises, scoping the extent of the economic promise in the lands they tracked. An American explorer, Henry Morton Stanley, was contracted by King Leopold II of Belgium to survey the possibility of a Belgian acquisition of the immense land of central Africa. Stanley had been turned down from undertaking a similar mission into central Africa by the British government, which was distrustful of Stanley and of the violent methods that many had witnessed him using on local African people. In 1889 Stanley set out on his first charge under Leopold, entering into central Africa from Tanzania on the eastern Indian coastline, to avoid discovery by other European powers – particularly the French, who would come to control Brazzaville in the West. In his letters to the Belgian monarch, Stanley would talk about the region's natural wealth, particularly of its rubber trees and teak wood. Initially Leopold's intention was to establish a large imperial territory in central Africa as a symbol of both his own wealth and power and the paternalistic civilizing policies of Belgian imperialism. It quickly became evident that the wealth of the Belgian Congo, as it was known until 1960, lay deep below the surface; and various colonial corporations began to establish themselves in the area, just as in the diamond mines of South Africa and what was then Southern Rhodesia (now Zimbabwe since 1980).

The most profitable mining company to be established in the DRC under Belgian rule was Union Minière. In his book on neocolonialism,[8] Kwame Nkrumah gives us one of the most detailed historical accounts of Union Minière's economic power in the country and of its role in the decolonization processes of the DRC. He writes:

In one year, Union Minière's profits were £27 million. But although the national production in Congo increased 60 per cent between 1950 and 1957, African buying power decreased

by 13 per cent. . . . The Congolese were taxed 280 million francs
to pay for European civil servants, 440 million francs for special
funds to Belgium, 1,329 million francs for the army.[9]

The DRC formally won independence from Belgium in
1960. Patrice Lumumba, who had been an important delegate
at Accra's All African People Conference in 1958, took office
as the country's first prime minister, a position he would hold
for just a few short months. In the country's independence
negotiations, Nkrumah narrates, various monies were to be
exchanged between the Belgian colonial administration and
the new independent government, including £120 million
portfolio shares of the Union Minière. The overall government
shares in mining companies that operated in DRC were valued
in 1958 at £267 million, but by 1960 – when discussions began
about their transfer to the newly independent government –
were deemed to be worth only £107 million, less than half their
value just two years before.

The handing over of these shares from the Belgians to the
newly formed Congolese government was conditional upon
political stability in the country. Moise Tshombe, who replaced
Lumumba as president, travelled to Brussels in 1965 to con-
clude the agreements and the transfer of money. Belgium,
however, demanded from the new government compensation for damage suffered by Belgian properties in the DRC
during the battle for independence, and also for the loss of
mineral resources supposedly owned by Union Minière and
other Belgian companies that operated in the region. Nkrumah
describes how the Belgian government insisted that the agreements also included a provision setting out that the new DRC
government must pay back to the shareholders, largely Belgian,
the interest that had accrued on the Union Minière bonds and
on which the new administration had supposedly defaulted. In
a speech delivered to the Ghana National Assembly in 1965, he
puts the following facts:

In the five years preceding independence, the net outflow of capital to Belgium alone was £464 million.

When Lumumba assumed power, so much capital was taken out of the Congo that there was a national deficit of £40 million.

Tshombe is now told the Congo has an external debt of £900 million. This is a completely arbitrary figure – it amounts to open exploitation based on naked colonialism. $900 million is supposed to be owed to United States and Belgian monopolies after they have raped the Congo of sums of £2,500 million, £464 million, and £40 million.[10]

So, with these concessions, the monies from the Union Minière portfolio shares slowly made their way back into Belgian hands, while a reduction in overall debt was diplomatically offered. Belgium announced that, of the £900 million debt, only £250 million had to be repaid. However, the Belgian government subsequently declared a further internal debt of £200 million to be repaid, together with a sum of £12.5 million in debt to be paid to Belgian private interests.

In total, despite the enormous wealth amassed by the colonialists from the DRC, the country was left with an impossible debt load from the unsteady moment of its independence, and the shares from Union Minière played a critical part in this bondage.

The company itself was hardly impacted. Indeed, it is common in decolonization processes that 'business interests' are protected, under the premise that economic gains are beneficial to everyone. 'Business interests', together with privately held property and land, were a major source of contention in the negotiations that aimed to end apartheid in South Africa. In the end, while government control was taken from the hands of the white minority, the country's colossal private wealth was largely untouched; properties and businesses held by the apartheid government were carefully privatized in the years prior to the first democratically held election in 1994. As mentioned in the previous chapter, Sizwe Mpofu-Walsh has written on the continued impact of the decisions made at the time of South Africa's independence from white minority rule. As for Union Minière, it continued operations until a merger in 2001, when it was renamed UMICORE, the first two letters, the website explains, in recognition of the company's history. Today UMICORE is one of the major multinational mining companies that extract raw materials and REEs for ICTs in the DRC, with renewed strategies for advancing their business in this area in response to the growing global demand for digital goods, of which AI represents the critical pinnacle.

The Electric Storm

As we look towards the future, the demand for raw earth minerals is set to increase significantly with the electrification of global society, which we know is a key foundation of AI. Promises of decarbonized economies and net zero carbon emissions come hand in hand with the electrification and digitalization of economic, social, and political life, since electricity replaces fossil fuels as the critical energy source. At the Paris Climate Accords global leaders committed to reducing carbon emissions by 50 per cent by 2030. As the pressure to meet the targets agreed to around this commitment intensifies, the efforts to support the transition from fossil fuels to electricity, together with other renewable energy sources, are set to increase. And yet a complex paradox lies at the heart of the decarbonization movement: the creation of electrical goods demands a higher and more intricate bundle of raw earth materials, and this causes other kinds of environmental and social harms.

Electric cars, for example, are sold as a high-tech climate-efficient vehicle, embedded in the latest AI technology – from self-driving capabilities to in-built voice assistance. Electric cars are increasingly becoming the vehicle of choice for environmentally conscious citizens who seek to make better decisions about how their consumption patterns contribute to global warming. Between 2020 and 2021 sales in electric cars have doubled. It is estimated that by 2040 almost all new car sales will be electric vehicles. Their electric batteries, like those in smartphones and laptops, are built using lithium; and 60 per cent of the world's lithium stores is found in the salt plains of South America.

In the far north of Chile, where the country borders with Peru to the north and with Bolivia and Argentina to the east, lies the Atacama Desert. Considered the driest place on earth, the Atacama Desert has on average less than four inches of rain per year. Even the clouds fail to visit, being blocked by the Pacific Ocean to the west and the high peaks of the Andes mountain range to the east. Over the millennia, nomadic indigenous groups have lived in the Atacama Desert and developed ways of life and cultural traditions, in syncopation with the rhythm of the desert's dry climate. Perhaps because of its scarcity, water is sacred for many of the indigenous groups that historically dwell in the Atacama region. For these communities, life is structured around the periods when water is available and the land gives vegetation. This careful and reverential relationship with the Atacama has been in some places threatened and in others completely destroyed by lithium mining in the region.

Over the past 10 years, the demand for lithium has increased by almost 1,000 per cent and, with that, its value. Statistics predict that, with the rising demand for AI-driven electric vehicles and other battery-powered devices such as laptops, tablets, and mobiles, the global demand for lithium will skyrocket. But mining it is not straightforward, especially in the arid salt plains of the Atacama Desert. Lithium is found within pools of brine suspended under the desert ground. These pools must be pumped out to the surface, and then filtered through evaporation processes to extract the precious lithium. Miles-long evaporation ponds stretch out across the desert, where, under the clear skies, evaporation happens quickly, refining the diluted brine into vivid technicolors of concentrated lithium. The brighter the hue of the pond – from yellows and greens to blues that reflect the desert's cloudless ceiling – the stronger the concentration of lithium. Often contamination can occur, releasing harmful toxins from the evaporation process into the ground and into groundwater.

These processes of extraction rob the arid landscape of the little water it has. For many of Chile's indigenous groups, the lithium industries have effected a kind of ethnocide, forcing them to relocate or to abandon their cultural traditions. In Chile, the rights of the country's indigenous population, which makes up at least 2 million of the country's total 18-million population, are notoriously unrecognized –both in the consti-tution of 1980, promulgated under General Augusto Pinochet's rule, and in a new draft constitution, which was rejected by Chileans in 2022. Without rights, these groups have little claim over the water now needed for the extraction of lithium.

It is not just local indigenous communities that are affected. Environmentalists have indicated that the desert's water table – the ground space between the surface and the water-saturated pools below – has been damaged, leaving the local wildlife to suffer. Without access to water, the population of Atacama flamingos is dangerously diminishing.

The benefits that the region as a whole gains from its lithium industries are limited. As in Africa, where the extraction of cobalt from the DRC begins a long and complex journey from supply chains outside the country to eventual use in the hardware of the digital world, in Chile, too, refining lithium for use in electric cars similarly involves a protracted supply chain where most of the profit-making takes place outside the place of origin – Atacama. Once extracted from the salt plains of

Latin America, the lithium is processed, refined, and turned into products for consumption in other parts of the world, where the requisite technology, infrastructure, and skills are available. China claims the largest market share in the production of lithium.

'Going green' has undoubtedly become an industry. This is not to say that going after the decarbonization of the economy is not essential to reducing climate-related risks; but it is to say that the process comes with its own risks. We must be alert to the harms and rising conflicts that the increasing demand for electric batteries will produce, and we must ensure that the effects – all the way down the supply chain – are taken into account in the policies and regulatory responses that are developed in order to govern justly the energy transition. Indeed, while electric energy offers an important alternative to fossil fuels, the environmental and social impacts of mining it are in some cases higher.

What must also be taken into account is the environmental impact of AI beyond that of its raw mineral supply chain. This is a particularly thorny issue, given the abstraction and complexity of AI's supply chains. There are limited data yet on the true environmental cost of AI and on which communities really pay the price and where. What we do know is that, as partly discussed in the previous chapter, data centres – the physicality of the cloud – require huge amounts of electricity to power them and thousands of litres to cool them. To train ChatGPT, 700,000 liters of water were needed to cool the whirring computers inside its mammoth data centre, an amount equivalent to the recommended daily water needs (which cover washing, cooking, and drinking) of 14,000 people. Recent research has also revealed that the carbon emissions of the cloud outweigh those of the entire airline industry, because of the energy required to power the relevant data centres.

We also know that the effects of climate change are unequally distributed around the world. There is increasing public recognition that the environmental damage caused by industries in the West and in other highly industrialized areas is felt in other parts of the world that already tend to be endangered, as the recent climate-related floods in Pakistan (2022) and Malawi (2023) indicate. Climate change is a global problem, but those whose activities contribute to it are not the same as those whose homes are flooded or whose food security is threatened because of it. As efforts are made to explore how AI can be used to help detect climate-related changes in the environment, it will also be important to find ways to concretely measure the environmental impact of AI through its supply chains and to demand greater climate-related accountability from tech companies.

AI sits at the very top of the digital food chain. In its ivory tower, it quickly forgets where it came from and who and what was involved in its making. Within the digital universe, no one is ready to take responsibility for the harm to people and societies that occurs right at the beginning or right at the end of the AI value chain. Laws and policies aimed at reducing the harms caused by AI fail to see the supply chains as part of the issue that needs addressing. It is always someone else's problem, largely because many others – other companies, other industries – benefit from, or use, the hardware that's created, or are also involved in dumping old electric waste. When everyone is involved and no one is to blame, there is no accountability and the harmful practices continue without respite. Also, everything is made more complex by the consolidation of economic and national power within these value chains. It is a power that will quickly override the uncodified claims of local indigenous groups, or will persuade a parent that finding cobalt is worth risking their child's life.

There are companies that have tried to map the value chains involved in the production of the goods they offer. Heineken sought to trace the lifecycle of their beer from barley to bar, in an effort to explore ways of doing better in every sphere in which the company was involved. Unilever has also made various efforts to track all the cogs in the supply chains that create its products. Its stated goal is to reduce carbon emissions, deforestation, and other activities within their sphere of influence that negatively impact the environment. It's not easy work and can in fact take years. In the case of Big Tech companies, whose supply and value chains are global, hugely complex, and multimodal, the work of tracing them may seem near to impossible. But if we really care to ensure that AI is equitable and contributes to a just and equal world, this work is necessary and important. It will require incentivizing Big Tech companies to respect human rights within their full value chains, which in itself is far from straightforward. Research tools such as indexes, which rank (and can name and shame) companies on the basis of performance in this regard, may be one such incentive. But such tracing is probably going to require a broader mindset shift in the way we think about the societal responsibility of Big Tech, and this is part of what we will work through in Chapter 7.

But the final piece is perhaps the most urgent: human labour. The human labour involved in AI and its production is a major piece of the puzzle of understanding AI's impact on global inequality. This is where we turn next.

4

The New Division of Labour

A young man enters a church. He is looking for a place to rest as he waits for the digital prompt that signals his next task. He works for a food delivery service, Deliveroo, cycling around the United Kingdom's city of Cambridge with an insulated box attached to his back, keeping the precious cargo ready for its receiver. The man's origins are North African. He came to the United Kingdom in search of a better world than the one he left, in which opportunities are scarce and dependencies on the young men in households are high. Despite finding work with Deliveroo, his wages are not enough for him to afford a place to live; he is homeless. During the winter months, he participates in a scheme run by a community of Cambridge churches to provide the city's homeless with refuge. Each day a different church becomes a night shelter, offering a warm meal and a camp bed under the cloisters for those who enter its doors. During his night in the church, he is intermittently hailed to complete delivery after delivery, and he returns after each trip.

* * *

Grace opens her phone, accepting a new request for work. She has registered with a platform that punts her skills to prospective clients, in return for a cut of her wages. She can offer house cleaning, childcare, and some light cooking, if appropriate. When offers of work come through, each job comes with a new house and a new madam to learn, navigate and respond to. Some treat her with kindness, some do not. She learns not to expect much. She makes about $10 a day, of which almost a half is used to cover her transport costs. Grace has migrated to Cape Town from a small village outside Harare in Zimbabwe, to find paid work. She leaves her young son in the care of her elderly mother. Taking him with her is unimaginable. How could she navigate finding work in the city with a small child in tow, and whom could she trust to look after him if and when she did find work? She works to send money back to her mother to ensure food for her son and, when the time comes, perhaps school too. Her family back home depends entirely on the money Grace sends back for their subsistence. She dreams of saving up enough money for a return bus ticket back up to Zimbabwe to visit her son. The bus ride takes about two days and will cost her around $100.

* * *

AI promises a world of ease. It will do the tasks and jobs we find tedious and dull, having an infinite appetite for the mind-numbing and an unfaltering aptitude for the repetitive. This idea has long persisted in the western history of robots and automata. Automata and, later, robots were often imagined in stories and crafted in workrooms as servants or soldiers, freeing up their masters to pursue higher, less worldly occupations. We can think of Da Vinci's Germanic knight, or Japanese tea-serving karakuri puppets. These conceits continue to play out in AI today. When Google, Amazon, and Apple began releasing their domestic or personal AI goods – Google Home,

Alexa, and Siri – they marketed them as digital agents that
would relieve their human owners of mundane tasks like
grocery shopping, managing diaries, or making travel arrangements. AI technologies are often depicted as aides to humans
who are in turn positioned as superior agents, fully in control
of the AIs at their disposal. This image of AI is particularly
important for maintaining its allure as a passive agent that can
bring us a life of leisure.

Equal to or Somewhat Better Than an Unskilled Human

For countries across the majority world, creating formal
employment opportunities is central to the manifesto of
almost every government. The eighth of the UN's Sustainable
Development Goals is to 'promote sustained, inclusive and
sustainable economic growth, full and productive employment and decent work for all'.[1] Without decent employment
opportunities and a living wage, the sustainable economic
growth needed to address global inequality is simply not possible. Creating opportunities to earn a decent, regularized wage
helps not only reduce income inequality; it helps address other
inequalities too, including in health and education.

Concerns about the impact that AI will have on employment
opportunities abound all over the world. Surveys measuring
public perceptions of and attitudes to AI provide important
barometer checks on what people think and feel about how AI
will affect their lives. (Such surveys are not available in sufficient numbers outside the United States and Western Europe,
and more research is certainly needed here.) A study in the
United States demonstrated a differential attitude towards
AI and automation between highly skilled college-educated
people and people working in lower-skilled jobs. While white-
collar workers believed AI to offer positive improvements to
workplace productivity and efficiency, blue-collar workers

were concerned about AI taking their jobs and livelihoods.[2] In South Africa, the only African country for which data on public perceptions of and attitudes to AI are available, 73 per cent of South African adults thought that AI would take over the majority of jobs within the next 10 years, and 60 per cent believed their job to be at risk from AI.[3] In many ways, answers to the question whether AI will assist in your job or will take it seem to mark the dividing line between privileged and underprivileged groups globally. The former believe that AI offers immense opportunities for wealth creation and remain somewhat concerned about killer robots taking over humanity in the future, while the latter worry about whether the AI-driven automation of today will take away their means of making a living.

This concern is not unfounded. Automation and digitalization have already brought about significant job losses. In banking, human bank tellers have been replaced by online banking. In fast-food restaurants, automated tellers are replacing counter servers. In factories, more and more tasks are being automated by hyperefficient robotics. With the rise of large language models (LLMs), huge numbers of jobs throughout the majority world may be lost. In my conversations with politicians across Africa, this is the biggest risk that the African continent and the majority world as a whole face as the empire of AI expands its reaches. Indeed, statistics from around the world demonstrate a potentially gloomy picture. A report from Goldman Sachs estimates that 300 million jobs are likely to be at risk globally from the new wave of generative AI.[4] More broadly, the report announces that up to two thirds of all jobs could be automated to some degree. These estimates are, however, extrapolated from data about the formal labour markets of the United States and Europe and have limited value for understanding what the impact of AI may be on the livelihoods of people who live in the majority world, where informal labour and economies are widespread.

One industry that effectively collapsed overnight when ChatGPT was released was essay writing. College students, in the United States or elsewhere, can pay to have their college essays written by someone at a fractional cost, maybe $50 for a 5,000-word essay. Many people across India and, indeed, other parts of the world used to make an income writing such essays. Yet, when ChatGPT exploded on the scene, it offered faster and cheaper essay writing (GPT-3 is freely available). It is unclear how those who made their living this way are finding work now, as their skillset in research synthesis becomes entirely superseded by generative AI models like ChatGPT.

As the capabilities of AI widen, particularly with advancements in natural language processing and the ability of machines to communicate with humans, the global division of labour is set to change. Globalization brought about the formalization of the global division of labour. As the world opens up through travel and communication technologies and industrialization is taking place across much of the majority world, particularly in the former colonies of European empires, multinational corporations lead a race to the bottom for cheap labour – a race that has begun during colonial times. In fact the old empires of Europe were built upon, and gave rise to, the multinational corporation – from the Dutch East India Company, the first joint-stock company in the world, which played a vital role in establishing a trade monopoly between South Africa and the West, to the mining companies we discussed in the previous chapter. During globalization and after the formal demise of European colonialism, the role of the former colonies in the global economy continued in the form of extraction of raw minerals and goods, expropriation of land for commercial gain, and exploitation of cheap labour. In return for foreign economic activity and trade, southern governments have worked to ensure a viable business climate for attracting investment (or are forced to do so through private

international law arrangements and foreign investment protocols) – often at the expense of their own people. Amazon's
Cape Town headquarters is a good example here, and we will
discuss it in a moment. But other examples are refraining from
implementing a minimum wage or certain labour standards
that may be less attractive to multinationals on the lookout for
the best (read: most exploitative) deal. We will examine this in
closer detail below, as we explore the kinds of work opportunities AI is creating.

In Asia, the industrialization at the end of the twentieth century and early twenty-first century was a central force behind
the boom of local economies, lifting as it did a number of
people out of extreme poverty. The Asia Development Bank
has reported that, between 1981 and 2015, those living in poverty across Asia decreased from 68 per cent to 7 per cent. Many
women were able to enter the formal job market, finding work
as seamstresses in textile factories and bringing wider benefits
to their families and offspring (although those industries were
not without their own levels of exploitation). With AI-driven
automation, including 3D printing, much of the manufacturing
labour that takes place across Asia may very well be onshored
back to the places where the products are sold and the profits
amassed, and where the cheapest labour will be provided by
AI. We will need to prepare for the significant gendered implications of these losses across much of the world.

The introduction of automation does not threaten just
the worker; whole industries across the global South could
collapse. As the capabilities of AI technologies increase and
specialize across different industries, there will be a decreas-
ing demand for cheap low-skilled labour from the majority
world. The capabilities of these kinds of AI systems rank low
on the taxonomy developed by Google DeepMind, which lists
the field's development towards artificial general intelligence
(AGI)[5] against a rising hierarchy of machine intelligence
(see Figure 5). At the top of the list is superintelligence, an

Generality → / Performance ↓	Narrow *clearly scoped task or set of tasks*	General *wide range of non-physical tasks, including metacognitive abilities like learning new skills*
Level 0: No AI	**Narrow Non-AI** calculator software; compiler	**General Non-AI** human-in-the-loop computing, e.g. Amazon Mechanical Turk
Level 1: Emerging *equal or somewhat better than an unskilled human*	**Emerging Narrow AI** GOFAI (good old-fashioned AI); simple rule-based systems, e.g. SHRDLU	**Emerging AGI** ChatGPT, Bard, Llama 2
Level 2: Competent *at least 50th percentile of skilled adults*	**Competent Narrow AI** toxicity detectors such as Jigsaw; smart speakers such as Siri, Alexa, or Google Assistant; visual question answering systems such as PaLI (Pathways Language and Image Model); Watson; state-of-the-art large language models for a subset of tasks (e.g. short essay writing, simple coding)	**Competent AGI** not yet achieved
Level 3: Expert *at least 90th percentile of skilled adults*	**Expert Narrow AI** spelling & grammar checkers such as Grammarly; generative image models such as Imagen or Dall-E 2	**Expert AGI** not yet achieved
Level 4: Virtuoso *at least 99th percentile of skilled adults*	**Virtuoso Narrow AI** Deep Blue, AlphaGo	**Virtuoso AGI** not yet achieved
Level 5: Superhuman *outperforms 100% of humans*	**Superhuman Narrow AI** AlphaFold, AlphaZero, StockFish	**Artificial Superintelligence (ASI)** not yet achieved

Figure 5 Google DeepMind's 'Levels of AGI' (2023)

intelligence universally superior to human intelligence, said to outperform 100 per cent of humans in any task. Near the bottom of the list is machine intelligence, which is considered to be equal to, or just a little better than, an unskilled human's.

This is where AI capabilities currently sit, in a rather crass admission that AI will render the 'unskilled human' economically deficient. For unskilled humans – many of whom are citizens of the majority world or belong in politically and economically marginalized groups in the West – the value of their labour is set to steadily decrease in the age of AI. In response to the narrowing of opportunities to work and earn a decent living, as people across the majority world in particular grapple with the global onslaught of a labour transition, desperation is exploited for profit, precarity reigns, and cycles of poverty and debt are further entrenched.

The policy response to this quandary of how to keep the unskilled employed in the age of AI has been to promote the idea of skills development. The broad strategy is that those whose jobs may be at risk should be offered opportunities to learn new skills, which would make them more attractive on a job market where they need to compete with AI. This is very easy to say and very difficult to enact. It assumes that workplaces have the resources and incentives to reskill at-risk workers; many won't. It assumes that workplaces are within the formal economy and that rules can be made that they will be obliged to follow; many aren't and won't (more on this below). It also assumes that people have the time and capacity to undergo reskilling. Especially for women in the majority world, who bear the burden of domestic and caregiving responsibilities that greatly reduce their available time, the idea of simply learning new skills for the world of AI is a problematic assumption, bound to leave many behind. This is exacerbated in regions where literacy levels are still low. In sub-Saharan Africa, approximately 25 per cent of men and a staggering 40 per cent of women are illiterate. In India, too, more men (81 per cent) than women (65 per cent) can read, particularly in rural areas. This contrasts with almost 100 per cent adult literacy in the so-called developed regions of the world.

The global gap in literacy and technical skills is a result of the historical forces of colonialism and, later, globalization, forces that sought to establish and maintain a low-skilled workforce that could be drawn upon to complete whatever rudimentary labour was needed. Discussing the history of skills development between the imperial centre and its colonies, Walter Rodney writes:

> Some hunted, some made clothes, some built houses, etc. But with colonialism, the capitalists determined what types of labor the workers should carry on in the world at large. Africans were to dig minerals out of the sub-soil, grow agricultural crops, collect natural products and perform a number of other odds and ends such as bicycle repairing. Inside of Europe, North America and Japan, workers would refine the minerals and the raw materials and make the bicycles. The international division of labor brought about by imperialism and colonialism ensured that there would be the maximum increase in the level of skills in the capitalist nations.[6]

What this ensured was that the profitability of raw minerals, together with the demand and development of the specialist skills needed to reap this profit, was concentrated in the imperial centres of power. We discussed this briefly in the previous chapter.

Critically, it is now these very domestic supplier industries in emerging economies that appear to be most at risk from AI-driven job loss, according to a report from the United Nations Industrial Development Organization.[7] While the report provides limited analysis as to why this is the case, the authors urge emerging economies to 'insert their industries into global value chains as suppliers to foreign industries that are increasing their adoption of robots'.

As an aside, where domestic supplier industries are on the decline and supply chains are increasingly global, the

environmental cost is higher and the ability to ensure transparency and accountability across supply chains becomes
significantly harder. Up until now, the increase in global supply
chains for the production and manufacturing of smart devices
and infrastructure has largely not been taken into account in
determining the real and potential environmental impact of
AI.

In the midst of these rising concerns about AI-driven job
losses, southern governments are under immense pressure to
sustain job markets and to create jobs. This comes at a cost and
governments face difficult trade-offs. I'll offer an example of
what these trade-offs can look like and whose rights become
the most easily tradable.

In Cape Town, near where I live, there is an area of land just
outside the city through which a river passes, flowing down
from the majestic Table Mountain. This land, Liesbeek, was
appropriated in the seventeenth century by Dutch colonists,
who removed the indigenous Khoi and San people from the
area and embarked on a series of farming and trade enterprises
that caused the loss of much of the local wildlife. Eventually
the land was taken over by the British, who, in time, established residential and commercial spaces on the river banks
and surrounding plains, constructing canal routes to manage
the overflow of water. Since the end of apartheid, the area
has been subject to protest actions, as the local city authorities permit new constructions, each one of which has various
environmental impacts and is in breach of national environmental impact assessment laws. However, in recent years a
development has arisen in Liesbeek that has caused the most
serious disagreement the area has seen in some time. Amazon
Web Services (AWS) began construction towards a massive
15.5-acre establishment that would be the company's regional
headquarters (Figure 6). From the local government's perspective, this new edifice will offer thousands of new jobs, helping
address a key priority policy area.

Figure 6   Amazon Development, Liesbeek, Cape Town
(photo taken by author)

A number of groups representing local traditional indigenous communities came together to oppose the development: the Goringhaicona Khoi Khoin Indigenous Traditional Council, the Observatory Civic Association, the Southern African Khoi and San Kingdom Council, the First Indigenous Nations of Southern Africa (FINSA), and the Kai Korana Transfrontier Royal House. Together they approached the provincial Western Cape High Court and, at a later stage, the national Supreme Court of Appeal, to challenge the development. Their case was against the Liesbeek Leisure Property Trust, which permitted the AWS development; and they cited environmental hazards and construction on historically significant land. The communities affected are aggrieved at not being meaningfully consulted about the use of land that historically belongs to them. Cape Town's colonial history is complex, and in the wake of the evils of apartheid the plight of the country's indigenous communities has been almost entirely clouded over. The political representation of South Africa's first nations communities is thin at best, with little acknowledgement of their struggles and rights in the founding constitution of the country's new democratic dispensation – which is not unlike the contempt for and lack of recognition of indigenous communities in the Chilean constitution (mentioned in the previous chapter). In addition, the legislation passed in 2019 to give recognition to the authority and customs of traditional communities – the Traditional and Khoi San Leadership Act – was declared invalid by the Constitutional Court of South Africa on account of the failure of the legislature to meaningfully facilitate public engagement in the development of the law.

Despite significant support for the campaign against the AWS development, the court application was dismissed in favour of the Liesbeek Leisure Property Trust. In fact, the development will generate an estimated 8,000 direct and 13,000 indirect jobs in a country where, after Covid-19, the official unemployment rate is at a remarkable 42.6 per cent. Against such numbers, the rights and heritage of indigenous communities, who represent a fraction of the population, became a sacrifice too easy to make.

Rationalizing Informality

When it comes to the majority world, the emphasis on AI's potential to disrupt jobs is misplaced, since most people make their living within the informal economy and are not, strictly speaking, in formal employment. Such 'jobs' are much harder to quantify, meaning that the current statistics and statistical models that have been developed to estimate how many jobs are at risk or will be created by AI are simply not relevant, as they pertain to formal economies and jobs. These models offer little guidance for the informal sector and, with it, for our understanding of the full scale of the impact that AI will have on southern job markets, livelihoods, and income inequality. The informal economy is defined in the negative vis-à-vis the formal economy, as casualized and unregulated economic activity where businesses and labour are unregistered and function without formal protection. Some 90 per cent of small businesses in the majority world are in the informal sector. Informal economies represent a historical culmination of colonialism and globalization, which caused a lack of state capacity to set, implement, and enforce market standards and low rates of education and technical skills across populations. Women make up a large majority of the informal workforce, although we should also note that much of the work performed by women the world over is unwaged. Jan Breman, a Dutch scholar who has researched and written extensively on these issues, defines informal economies thus:

> The whole gamut of economic activity consisting of petty business, low capital intensity, low productivity, inferior technology, reliance on family labour and property, skill formation mainly by training-while-doing and finally, a small and poor clientele.[8]

In Africa, the vast majority (around 90 per cent) of people earn a living this way. Work may include selling food or beverages, transporting goods, people, and animals, domestic work and gardening, serving as cleaners and farm workers. Those who perform this kind of work tend to be land-poor or landless, without ownership of any means of production, and able to make a living only through the labour they can offer. Many are migrant labourers who come to wealthier countries or cities in search of work opportunities, like the migrant labourers who come from the north of Ghana to Agbogbloshie in search of work – as we saw in the previous chapter. Today many of these people take work as Uber drivers or food delivery workers. Throughout the world, there is an overrepresentation of migrant workers in platform economies – workers like the young man described at the beginning of this chapter, who works for Deliveroo. In South Africa many people from rural parts of the country migrate to the city, in search of work, and leave their children to be raised by grandparents or older family members. Typically, women find work as domestic workers and may well sign up to platforms like SweepSouth. Their stories are generally similar to that of Grace at the start of this chapter.

In the age of AI, these new platform economies have arisen as a new economic model, premised on the idea of value co-creation between workers, customers, and the platform. Where customers are concerned, they do get value for money, being able to choose the best offer from a selection of service providers. As for workers, these platforms supposedly offer them flexible employment 'gigs'. The World Bank estimates that there are 435 million gig workers globally, which makes up almost an eighth of the global workforce. Estimates calculate that there are at least 30–40 million gig economy workers across the global South alone. By 2025, platform economies are set to make up 30 per cent of global economy activity.

While these platforms certainly offer opportunities to make cash, such opportunities come with a catch. Legally, platform workers occupy the grey space between being an employee and being self-employed, but without enjoying the benefits of being either. If they were formally self-employed, platform workers would be able (in principle) to determine their own wage and hours of working, but they don't. If they were formally employees, they would be entitled to benefits, but they aren't. Workers operate with limited job security, such as insurance, healthcare, or leave pay. The price of the convenience of having food delivered or having a taxi always at hand is paid by these workers. And, when the demand is not there, the workers are not paid. Uber drivers, for example, are not paid for the time in which they wait to be hailed. This can mean driving around in high-demand places and burning through petrol that can't be reclaimed, or parking in locations where it is free to park but further away from potential riders.

Because of the lack of formal labour protections, when Covid-19 hit, many gig workers were left without any kind of income. In places across the majority world where unemployment rates pose critical challenges to local development, what is needed is long-term and secure jobs – reliable full-time employment, not touch-and-go gig work where a salary is not guaranteed and labour protections are limited at best. Labour protections might include unemployment insurance, protection in case of occupational injuries or disabilities, as well as collective bargaining rights. Without these protections, platform and gig economies effectively rationalize the informal economy, squeezing value and profit from its workers, while denying meaningful labour rights or the opportunity for full-time salaried employment.

Poor labour protections are not the only unethical aspect of gig work. If gig workers hoped for support and understanding from their managers, they'd be wrong. Platform managers are clinical and exacting. The reason: they are AIs. Gig workers are not managed by humans who have the capacity to offer compassion and understanding, but by an AI programmed to drive out efficiency and profit and to maximize the value that can be extracted from a worker and their time. Workers in the gig economy are routinely subject to intrusive surveillance methods that monitor their every movement and clicks, and to the nudging and control of behaviour exercised by algorithmic systems that seek to optimize performance. This creates an enormous asymmetry of knowledge and imbalance of power between the gig worker and the omniscient AI system. In a job where the boss is always right and cannot be questioned except through an automated decision tree of pre-programmed questions and answers, work becomes a ceaseless experience of AI-powered gaslighting.

Again, we see how the price of a world of comfort and ease for some, as food is delivered to our door and taxis are readily available, is paid by the underemployed – those whose work is not enough to make them meet their needs and live a dignified life. Just as in the stories of Amazon factory workers who are compelled to wear adult diapers so as to ensure that no minute is lost in the retrieval of items ordered online, much of the dignity of work is sacrificed when an AI is your boss.

Platform gig work is increasingly becoming one of the most readily accessible work options for many people around the world, both in richer and in poorer countries. For many gig workers, this is one of several jobs that they usually hold down in order to make ends meet. With rising inequality *within* countries, especially in the global North, and with layoffs, which are often due to automation, more and more people are being driven to earn a living or supplement their income through various kinds of digital gig work. But what this offers in terms of so-called flexible employment does not come anywhere near meeting the standards for the kind of formal employment that is needed both for socioeconomic mobility and to lift people out of poverty.

Efforts to formalize informal economies constitute an integral aspect of poverty reduction strategies. Central to these efforts is to ensure that workers have access to labour protections and to mechanisms for collective bargaining, such as trade and worker unions with which to negotiate fairer wages. However, historically, these informal structures have been what allowed poorer countries to participate in global markets, offering cheap, largely unregulated labour. For southern governments, establishing decent labour standards and protections – which includes implementing minimum wage measures – meant risking a loss to foreign investment, as companies seek more competitive employment rates elsewhere. As governments and societies everywhere grapple with the changing divisions of labour, if inequality is to be addressed and not entrenched, then ensuring strong labour protections, and especially measures for collective action, should be a priority area for regulatory and policy reform.

I Am Not a Robot

Just as the race to the bottom prompted multinational corporations to seek out cheap labour in poorer countries of the majority world, the labour demands of AI are being offshored to places out of reach, where the supply chain that leads back to Silicon Valley's AI developers is harder to trace, and the commitment or investment is expected to be small. It is an industry that is kept abstract, informal, and unconnected to the larger AI ecosystem. In contrast to the high income-earning data scientists of Silicon Valley, the labour force required to do the tedious work of building AI models toils in unseen places, plunged into regimes of work that cannot sustain a decent standard of living and trapped in cycles of overwork, debt, and varying degrees of maltreatment and harm.

In recent years the painstaking human labour involved in the production and maintenance of AI systems has been brought to light. Stories reported in mainstream media have revealed the massive human labour force that has quietly arisen in hidden places around the world, tasked with painstakingly building – and rebuilding – the digital archives and data vaults behind AI.

As we discussed in Chapter 2, AI is built from great swaths of data that are used to train and refine its models and systems. With the advent of generative AI and LLMs, the demand for data has reached new heights. One of the most sophisticated LLMs released to date has been Microsoft's DaVinci. With OpenAI's GPT-3 as its base, DaVinci has an enormous parameter count of 175 billion; this refers to all the variables within the model's neural network, which is trained on 45 terabytes (i.e. 45 trillion bytes). The value of these systems rests almost entirely on the degree of accuracy of the results they produce when prompted with input data such as: 'Write an essay on the role of the Council of Nicaea in the Christian church.' This requires not only volumes of data but a complex set of interrelations between a machine and a human who teaches the machine precisely how to interpret the data of the world to which it is exposed.

Data do not arrive ready for AI's consumption. They must be organized, labelled, and structured – a kind of labour that, at present, only humans can do. Once an AI model has been trained on an initial set of data, there is another layer of human labour that is needed to adjust and correct its outputs, ensuring that it interprets and feeds back to the human world correctly. In computer speak, this is called reinforcement learning, or reinforcement learning from human feedback (RLHF).

Any Internet user will be familiar with simple exercises whereby entry to a webpage requires proving 'I am not a robot' by identifying, for instance, a series of pictures of motorbikes. Imagine spending 12 hours a day (or more), for months on end, completing such activities. This is precisely the kind of work that is involved in the data maintenance needed to build and then boost the accuracy and efficacy of AI systems. The work is not easy. It is exacting, yet mind-numbing and thankless work, and more so given the wider set of problems that surround current labour practices in this space. The kind of pay that is being offered to those involved in the sorting and labelling of data for AI in the South is both way below a living wage and far lower than what tech companies would need to pay workers in their home country to do the same job. This kind of work often involves viewing thousands of violent and disturbing online images and videos and removing them from mainstream social media platforms.

Thanks almost entirely to some individual whistleblowers such as Daniel Motaung, accredited by *Time* magazine, the appalling labour conditions of this growing industry of digital microwork have been exposed. In his own words, Daniel Motaung describes his experiences as he worked at a content moderator firm in Kenya, where he was tasked with moderating extreme content for Meta:

The very first video I watched in my new role was a live video of someone being beheaded. The work of content moderators involves long shifts of watching a constant stream of graphic violence, sexual abuse, animal torture, and sexual exploitation of children.

Nobody survives this work unscathed. Several of my co-workers were diagnosed with post-traumatic stress disorder (PTSD). I feel like I am now living in a horror movie. The trauma I was subjected to keeps replaying in my mind. In the flashbacks I am often the victim of the violent content I had to watch.

Before I joined Sama I had a pre-existing condition of epilepsy, which was inactive in 2018. In other words, it did not affect my life or need treatment. But during and after my time as a Facebook moderator, the seizures returned. I believe this was a result of the trauma and ill-treatment.[9]

Digital workers are recruited by pop-up companies, then contracted by firms that have themselves been contracted by an AI lab to curate the data needed to feed the growing AI. These workers are set up in temporary office spaces, on short-term gigs. Their work is piecemeal and abstract. In an article published by the *New York* magazine and the *Verge* and titled 'Inside the AI Factory', Josh Dzieza describes how 'each project was such a small component of some larger process that it was difficult to say what they were actually training AI to do. Nor did the names of the projects offer any clues: Crab Generation, Whale Segment, Woodland Gyro, and Pillbox Bratwurst.'[10]

But the effect on the individuals who performed this work is devastating. Daniel Motuang reported that no support was offered to workers who processed emotionally distressing content. Another worker reported suicidal thoughts experienced after days of repeatedly witnessing heinous crimes. In his article, Dzieza emphasizes how the work was 'stripped of all its normal trappings: a schedule, colleagues, knowledge of what they were working on or whom they were working for. In fact, they rarely called it work at all – just "tasking".' They were 'taskers'. The work of taskers is not paid per hour or minute but per task, and some tasks require hours and hours of unpaid research to be undertaken before starting. The word 'task' comes from the Latin verb *taxare* (to peck, censure, appraise): not only does it designate a job to be done, but it also implies that the job in question *has* to be done. A task is a piece of work or a quantity of labour that has been imposed as a duty. In this light, the kind of terminology that surrounds the work of AI's data labourers may be deeply problematic, suggesting servitude to an imperial industry.

What is more, a recent investigation into digital labour in the majority world revealed how refugee camps have become a new point of exploitation for Big Tech and other platforms that seek out cheap digital labour. Dadaab, one of the world's largest refugee camps, is located near the eastern border of Kenya and shelters a quarter of a million refugees from across the African continent. Here tents containing hundreds of computers and yards of wires have been erected to facilitate click work undertaken by the camp's residents. Residents are offered 'work' opportunities, but, the report indicates, rather than being paid in real money they are paid in tokens for use in the camp, thus facilitating only the continued circulation of the camp's closed economy. A similar report, also published, relays how a LLM company in Finland is paying prisoners in Finnish prisons $1.41 an hour to help train their chatbot. This labour can't be exported to the global South; instead, Finnish-speaking vulnerable people need to be found and exploited.

It is abhorrent to imagine that people who are already vulnerable as a result of their experience of war and forced migration are the very same people tasked with curating the worst of the Internet's violent content. Their vulnerability is exploited to ensure that the rest of us don't need to witness this content online. It is as if suffering were contained where we can't see it – whether offshore or in closed-off places. Indeed, we were not supposed to see it. The only reason why these issues have been brought to light is the bravery of a small number of whistleblowers and investigators.

Workers, Resist!

AI's underclass is deliberately depoliticized. Its work is isolated and isolating; it consists in performing tasks remotely and online, or delivering goods and transporting passengers while guided only by an automated app. Workers are isolated from the everyday human contact that makes work both socially and politically meaningful; the connection, such as it is, between their work and the world is mediated by an indifferent AI system. Starved of this connection, it is almost impossible to resist the machine, to challenge exploitative practices, and to demand better working conditions and labour rights that lay the foundation of dignified work. This is exacerbated when AI's labourers are forced to work in the grey areas of the law, without formal labour protections or unions, which usually provide some balance to the power asymmetry between employees and their employers. If the capacity for collective action from AI's workers is restricted, AI's power can go unchecked, which can lead to all sorts of inequitable and volatile outcomes.

The argument goes that there is an inevitable diffusion of technological advancements and benefits across the world. It is, they say, only a matter of time. The reason why we aren't there yet, or why the so-called developing world has not been able to fully take advantage of earlier digital technologies, economic historians insist, is an inability to use labour effectively.[11] These statements fail to take into account the history of the acquisition of indigenous technical skills through colonial exploits, or the decades of market fundamentalism that drove southern governments to prioritize foreign investment over fair labour standards and protections. With the shiny economic promise of AI, we are poised again to see a deprioritization of labour rights in the face of the potential economic gains of AI. When we look at the promise of economic growth from AI, we have to reckon with the question of who, and whose labour, paid the price for this.

The political power of AI's working class is centrally important in driving forward the kind of equitable and just world we want AI to serve. These are the people who really should be at the centre of debates and discussions around AI and how to regulate it. And protecting their rights should be a priority for governments everywhere. This would require governments to fast-track the development of fair labour rights and protections that extend to contractors and to gig and platform workers. Such an action would presuppose the establishment of a minimum wage and the existence of institutional structures that can enforce these standards and protections, as well as protections and opportunities for collective action and bargaining power. In places across the majority world, while there is much to be done, positive steps are being taken. In Rajasthan in India, for example, a new Gig Workers Act has been passed, providing welfare and decent working condition protections for gig workers in the state.

While resistance to unfair labour conditions should be formalized through labour protections and rights, there are, of course, many ways to resist – some more playful than others. I once took an Uber ride with a young man called Hussain. Hussain had come down to South Africa from Somalia, in search of work. He was young, bright and determined not to take the strictures of his electronic boss too seriously. As we drove through the streets of Cape Town, a high-pitched helium-filled voice instructed the route home. Hussain had altered the settings on his AI-powered voice assistant so it spoke at the highest pitch, turning the neutral flat timbre of the computer-simulated voice into an irony-filled *reductio ad absurdum*. It felt like a tiny triumph in the age of AI. And we laughed all the way home.

5

Fit for What Purpose?

The humidity drips off the wall of my hotel room. I look out over the rushing ocean coast of Dakar, the capital city of West Africa's French-speaking powerhouse, Senegal. In the distance, the Island of Gorée rises up on the horizon. Now a UNESCO heritage site, the island was once a major trading post for the French colonial empire. Of all that was traded at Gorée, slaves were the most significant. Many hundreds of thousands (if not millions) of people originating from various parts of West Africa spent their last days and months in Africa on the Island of Gorée, housed in the Maison des Esclaves. In a basement chamber of the building, there is a small stone window framing a view of the Atlantic Ocean as it spreads out endlessly towards America. After visiting the Island, the Canadian poet and novelist Dionne Brand wrote of the window as the Door of No Return, marking a fissure in the historical identity of the men and women who endured the crossing of the Atlantic Sea in the fifteenth and sixteenth centuries and of their contemporary descendants in search of their heritage.

As I look out over the vista that leads to the Island of Gorée, I am preparing to speak on a panel discussion the next morning, a panel that – I am acutely aware – is one of the most significant panels I've been invited to speak on to date. The other panellists include Juliana Rotich, who founded Kenya's first innovation hub and who heads up Kenya's famous digital mobile money platform, M-Pesa; David Sengeh, Harvard University graduate, celebrated author of *Radical Inclusion: Seven Steps to Help You Create a More Just Workplace, Home and World*, and Sierra Leone's youngest chief minister and chief innovation officer; Sabin Nsanzimana, Rwanda's minister of health and the man behind the country's containment of the Covid-19 virus; and Bill Gates, speaking in his capacity as co-chair of the Bill and Melinda Gates Foundation.

Panels like this one take some preparation. In the run of the show, the question posed to me was: 'Can AI redefine the lives of the poor?'. It is – no pun intended – a billion-dollar question. In a radio interview later that month, I said that it was a question that would be likely to preoccupy me for the next few years of my career. The high rhetoric of AI pioneers like Elon Musk and Demis Hassabis asserts that AI will be 'the solution' to poverty. But these easily spoken statements elide how deeply complex and multidimensional poverty is, and that it is inseparable from inequality. Women, on average, constitute the poorer sex; people of colour tend to have less wealth than white people; people with disabilities have a harder time securing work; and countries in the majority world that were colonized by Europe have far less national wealth than their previous colonizers or the lands of settler colonialism in North America. No one solution, not even the supposed 'meta-solution' of AI, can address the multi-crisis that is poverty, together with the historical and structural factors that have produced its form today.

Putting aside for a moment whether the potential of AI and of the advanced AI tools that have been developed is pure rhetoric or real possibility, to translate this potential into actions that meaningfully and sustainably transform the lives of people who currently live in conditions of destitution will require an enormous, coordinated, global effort not just from governments, international bodies, and Big Tech but from people everywhere. Part of the reason for this is the scale at which the problem sits and the scale at which AI works (which we will explore in greater detail in the next chapter). Everyone is already impacted and involved, but some benefit enormously while many others lose out. The dividing line between the winners and the losers of AI is gendered, racialized, and locational. But the responsibility to look after those who bear the costs of AI falls on those who reap its benefits. Those who benefit can influence considerably whether or not AI is put to work as a transformative poverty eradication tool, because these are the people for whom AI is currently being designed and built. Where AI is designed to generate profit and entertainment only for the already privileged, it will not be effective in addressing the conditions of poverty and in changing the lives of groups that are marginalized from the consumer markets of AI. Deep changes are required if AI is to change the lives of the poor for the better. The changes needed are complex and reach far beyond the design and use of AI itself. They require developing technology that actually meets the needs of those who live in poverty around the world and, critically, this means changing the incentives that currently drive the growth of the AI industry. This was the point I ended up making at the panel discussion in Dakar: that the incentives need to change if we are ever going to have an AI that is truly transformative of the lives of the poor.

In this chapter we will examine how people who live in poverty are directly affected by AI. We will do so by looking in particular at AI-driven social assistance programs and at the use of biometric technologies, which are fast becoming a core area of digital development for governments across the majority world. As in the case of the rise of gig work with limited labour protections (discussed in the previous chapter), here too AI contributes to the fact that too many live in social and economic precarity. We will also come to understand how this is worsened by the impacts of these technologies on social and community cohesion.

AI systems are increasingly deployed to manage the situation of people who live in poverty. Some may have every good intention behind them; others are downright punitive. In a book titled *Automating Inequality: How High Tech Tools Profile, Police and Punish the Poor*, Virginia Eubanks, a leading American political scientist, has written about how data-driven technologies target those who live in poverty in the United States. Her work clearly demonstrates how AI can function as a poverty trap, locking people into cycles of debt that become almost impossible to break. For example, AI systems are now used to determine whether an individual gets access to a loan from the bank, or – in many parts of Africa, for instance in Nigeria and Kenya – to microcredit. These financial services are often a crucial lifeline for people who live in poverty; but, as AI technologies are integrated into the decision-making process behind who gets such access and who does not, their value for these people may dwindle. They are designed, after all, to generate a higher level of accuracy in the risk assessment associated with the underwriting of a loan – not to ensure that more people have more access to financial services.

What remains fairly constant across the suite of technologies developed for the poor and, more broadly, for the field as a whole is that the people who may be most negatively affected by the outcomes of an AI system are typically not consulted or engaged in its planning and design. This is exacerbated in countries across the majority world, where people are even at a further remove from technology hubs like Silicon Valley, and where their experiences and realities are barely reflected in the datasets upon which AI models are developed and trained – let alone in how the user interface or experience of the technologies is designed. If AI systems are not from the beginning developed to be inclusive of people outside western paradigms, they are at best not fit for purpose and, at worst, they deepen inequality.

Blood Money

In June 2023 Human Rights Watch (HRW) released a report detailing the harms caused by a new AI-driven algorithmic technology used by the Jordanian government to provide cash transfers to the poor.[1] The system was named Takaful, in a nod to the traditional Islamic notion of social welfare, which goes by the same name. Cash transfers are a means of non-contributory social assistance (that is, they are not tied to contributions made by an individual or employee, such as tax or employment insurance) and have been evidenced to be an important measure to address poverty, but are often reserved for the poorest in society. There has been much debate about the benefits of cash transfers, particularly whether or not they provide a means for people to make their own decisions about where money is most needed. For some, cash transfers offer opportunities for financial agency and independence; for others, they are simply not enough to cover the range of legitimate expenses a family may have. Worse still, where cash transfers are paid out to a family, research has demonstrated that often the money is squandered by the male head of the family on gambling, on drink, or in other ways.[2] But, overall, where cash transfers have been fairly liberally available, they can offer a critical lifeline for indigent groups, reducing poverty across countries. This has been the case with the celebrated Programa Bolsa Família, a family cash transfer system (also supported by the World Bank) that was introduced in Brazil in 2003 and brought cash relief to over one fifth of the country's population.

AI has been touted as an effective new tool for determining who in society is most in need of financial assistance from the state, and for distributing transfers accordingly. Endorsed and funded by the World Bank, the Jordanian AI system investigated by HRW was by no means unique to the country but is part of a suite of AI-driven social protection initiatives that are rolled out by the group in the Middle East and North Africa region.

HRW's reporting set out how the AI model used in Jordan collected data from various sources – such as household bills for water and electricity, car registration and licences, participation in benefit schemes, employment history and records – and assessed them, together with the answers provided by applicants on a mandatory living expense form they had to complete. Takaful collected data against 67 socioeconomic indicators in order to create a ranking of Jordan's poor, thus establishing an intimate picture of the lives of those most in need. But its aim was not to understand who was living in poverty; it was to determine who was really the most deserving of the state's assistance. This resulted in a complex scoring and ranking matrix, which downgraded people who had owned a car for less than five years and automatically excluded everyone who did not fall under the outdated figure for what constituted the Jordanian poverty line, even if they had far less than needed to meet their socioeconomic needs. In addition, those whose households were not headed by a Jordanian citizen were counted as a family of one person, no matter how many members they included. In many cases, the mandatory expenses form could not be submitted unless the applicant engineered their monthly expenses to be equivalent to their monthly income. In its report, HRW captured stories of people who, living in rural areas, had to spend more than $6 just to travel to a mobile phone shop to withdraw cash. In all, Takaful provided cash transfers of between 40 and 136 Jordanian dinars, the equivalent of $56 to $192, to the households it deemed eligible. This was nowhere near enough to fill the gap.

In May 2023, just before the report was released, Jordan lifted the moratorium that had allowed relief from the country's draconian criminal laws on debt. For those unable to repay a debt burden, imprisonment was a very real possibility. In an Islamic country like Jordan, the concept of 'debt' is determined by shariah law and can take a number of forms. Indeed, a 'debt' may not always be a bondage of money owed; it can be instead a debt owed to a family for a life taken. According to Islamic doctrine, the next of kin of the deceased can claim from the accused blood money (*diyah*) as indemnity. As societies modernized, social insurance schemes were developed at the community and tribal level to ensure that a mass of compensation was available in cases of murder where the killer was unable to pay the *diyah* themselves. Takaful – the name adopted by the AI-assisted World Bank-led scheme in Jordan – is the traditional Islamic name of a wider social assistance program of this kind within shariah law that collects money from members of society in order to provide for contingencies and for other members of society who may, for whatever reason, suffer economically. Traditionally, this is a social and collective scheme that seeks to ensure the wellbeing of society. It is built on principles of social cooperation and shared responsibility. Losses and liabilities are shared among all, and the goal is to ensure that no one person can gain advantage over others through the system.

This stands in stark contrast to the World Bank's AI-driven Takaful. Where the proper Islamic concept corresponding to this name (*takaful* 'solidarity') seeks to build community solidarity, trust, and compassion, HRW reported how the AI-assisted Takaful drove divides between community members who were competing for the meagre resources it made available. 'Its formula', the report reads, 'flattens the economic complexity of people's lives into a crude ranking that pits one household against another, fueling social tension and perceptions of unfairness.'[3] What is not clear from the policies and reports of the government of Jordan and the World Bank is why the name Takaful was chosen for the AI system: was it deliberately set up as a new AI-driven mechanism for advancing traditional Islamic *takaful*, or was it a marketing tool to assure individuals that this was a legitimate, shariah-compliant scheme set up to benefit them? In the official documentation about the program, the word *takaful* frequently appears in inverted commas, as if to demonstrate that the concept is being borrowed from a context that does not quite align with the new way in which it is used. In addition, there are no references to shariah practices and values in the description of the program's rationale, design, or evaluation, which suggests that the system had little to do with the proper Islamic meaning of *takaful*.

Biometric Empires

Jordan is by no means unique in its approach to using AI to determine and administer citizen benefits. Governments throughout the world are increasingly adopting AI solutions to power more effective social protection regimes, where the validity of every state benefits claim can be individually assessed against risk factors and regression curves. In the United Kingdom, AI techniques are being incorporated into the Department of Work and Pensions' communications systems for benefits claims and disputes. The origins and owners of the AI systems that are deployed in such cases tell an equally disturbing story of inequality: while the poorest segments of the British population struggle to reason with welfare robots, the AI company behind these systems – UiPath – is co-founded by Daniel Dines, reportedly the world's first bot billionaire. With AI, poverty is big business.

Across Latin America, where the implementation of social protection policies that grant cash transfers, pensions, and school meals has been integral to the region's progress in addressing poverty, AI is now used to drive further efficiencies within these systems. Laid on top of biometrics and the digitalization of critical government services, AI offers the state a highly effective tool with which to manage its population centrally. Biometrics is a technique of identification and authentication based on the patterning of unique physical, psychological, or behavioural human markers – from fingerprints to face, gait, voice, emotion, and even heartbeat recognition – and is increasingly becoming an important domain for government use of AI processing all around the world.

For so-called developing countries, rolling out biometric digital ID systems (bioIDs) is considered a key enabler – an instrument that allows a digital society to tap into the economic potential of AI and supposedly facilitates citizens' access to government services. India's tech stack is a prime example of this. 'Technology for 1.2 billion Indians',[4] the India Stack is a government-led initiative for a fully integrated national digital ecosystem. At the base of the stack is the biometric ID system Aadhaar (the name comes from a Hindi word meaning 'base' or 'foundation'). Aadhaar is required for individuals to access any kind of state benefit; but it is also needed for voting. Within the India Stack, payment systems (such as mobile money) and personal government-held data (certificates, driving licences, etc.) are layered on top of Aadhaar, enabling a network of application programming interfaces (APIs) that can facilitate seamless transactions between people, businesses, and government. Almost the entire adult population in India is now registered with the Aadhaar, which makes this the most ambitious and populous digital ID system in the world.

A whole host of concerns have been levelled against the Indian Aadhaar, from state-led surveillance and discrimination to the on-selling of personal data without consent. Through biometrics linked with AI, which systems like Aadhaar make possible, an individual can be continually tracked and monitored, providing the state (and other actors) with intimate and sensitive real-time information about their daily life, often with devastating effects for people who do not behave as expected or do not fit in with the norm. What is so problematic about biometric technologies is how invasive they are on the human body, coupled with the increasingly imaginative uses to which they are put by groups in power. Their use creates an enormous power imbalance between individuals, who cannot defend the seemingly incontestable evidence about their biological being, and the owners of the biometric technology, be they a state-led security group at border crossings[5] or a private company seeking to optimize workplace surveillance on its employees. Think of your smart watch being controlled by your employer, who was monitoring your focus at work and your emotional reactions to new policies. Or think of the film *Minority Report*, where the police use mind-reading technologies to predict criminal behaviour before it happens, and this becomes an omnipresent aspect of social and public life.

Access Now, a global digital rights campaign group, in its report on biometric technologies and human rights, highlights how the technology behind biometric systems is opaque and unexplainable, impacting on people's rights to challenge these systems, or even to know when they are being used.[6] A concerning finding of the report is around the function creep associated with biometrics. The authors note how 'the biometric market is highly lucrative, with fierce competition and pressure to constantly patent new technologies – resulting in outsize industry claims around expertise and what technology can do'. With huge market incentives to expand the reach of biometrics and with such value placed on datasets containing biometric information, the conditions are ripe for exploitation at the cost of individual rights. Examples offered in the report include a biometric voice recognition system that claims 'to detect mental distress or anxiety markers, or an eye-tracking tool that tracks attention and "nervousness" for use in a clinical setting'; and these 'can then be repurposed in "AI lie detectors" used by law enforcement and the military'.[7]

One of the most shocking recent examples of this kind of use took place in Kenya, with a technology known as Worldcoin. Touted as a 'more human' passport for the Internet and as offering universal proof of its humanness, Worldcoin claims to be an open-source protocol designed to give everyone access to the global economy. More specifically, it is a cryptocurrency platform that offers cryptocurrency to those who give the company their iris biodata. The iris is considered the most accurate biodata for personal identification: it is entirely unique – even twins have very distinct irises; it does not change over the course of an individual's lifespan; and its sui generis structure and makeup contain more data points for comparison than do the face and the fingerprints. Rather more ominously, the iris is also harder to disguise. When authorities in Hong Kong were using AI-driven facial recognition technologies to identify protesters, many of them took to wearing various types of masks for subterfuge. We heard, too, in Chapter 1 of the migrants who burnt their fingertips to claim back some agency over the fingerprinting identification systems to which they were subject. But an iris is harder to cover up, and much harder still to self-mutilate.

With the power of iris biodata, Worldcoin set out to create nothing less than a new global currency available to everyone – at the cost of their biometric data. Founded by none other than Sam Altman, the founder and CEO of OpenAI, the company has been mired in scandal from the beginning. An investigation from the Massachusetts Institute of Technology (MIT) found that, to recruit the company's first half a million users, targeted campaigns were carried out largely in the majority world and in low-income communities across the West. People were manipulated, deceived, and bribed to sign up. The report describes these campaigns:

In villages across West Java, Indonesia – as well as college cam-
puses, metro stops, markets, and urban centers in two dozen
countries, most of them in the developing world – Worldcoin
representatives were showing up for a day or two and collecting
biometric data. In return they were known to offer everything
from free cash (often local currency as well as Worldcoin
tokens) to Airpods to promises of future wealth. In some cases
they also made payments to local government officials. What
they were not providing was much information on their real
intentions.[8]

Although concerns were raised, the company continued to
receive investment and to induce new users to sign up. In
August 2023, the Kenyan authorities issued a decree to sus-
pend Worldcoin activities in the country; the country where
Worldcoin had first launched and the country with the largest number of sign-ups, estimated at around 350,000, was
Kenya. Here Worldcoin was offering around $50 of cryptocur-
rency to anyone who would sign up, taking their iris scans in
return.

Another example is that of the Chinese facial recognition
company CloudWalk approaching the Zimbabwean government in 2018 with a deal. CloudWalk would install surveillance
and facial recognition infrastructure in Zimbabwe and would
invest in the country's digital ID system, which was being
advertised as the key mechanism for ensuring a free, fair, and
safe election later that year. In return, the company would get
access to the facial biometric data of the Zimbabwean national
population registry, which holds picture records of the faces of
millions of Zimbabweans. The deal was signed a few months
before the first national election since the passing of Robert
Mugabe, who had been the country's increasingly despotic
president after independence, more than 30 years earlier. For
CloudWalk, the opportunity arose at a time when the issues of
racial bias in AI were peaking and pressure was mounting to

improve the accuracy of these models. The Zimbabwean facial recognition data were a goldmine for CloudWalk, providing the company with millions of African faces to augment the training data of its models.[9]

Zimbabweans knew little about what was taking place. No consent for the sharing of their intimate personal data was obtained. In fact, at the time, Zimbabwe did not have a data protection law that might have rendered the data transfer illegal or provided Zimbabweans with recourse for wrong-doing. The digital ID system that was to assist in the election process was hamstrung for various technical and bureaucratic reasons, ultimately creating distrust around both the election process and the use of digital biometrics more broadly. In all, the entente between CloudWalk and the Zimbabwean gov-ernment exploited the people of one of the world's poorest countries to boost the profits of China's biggest facial recogni-tion AI companies.

Keith Breckenridge, a South African scholar who has scru-pulously studied the history of biometrics – particularly in the making of the South African apartheid state and in the country's relationship to India's independence movement – defines biometrics as 'the identification of people by machines'.[10] In his book, Breckenridge explains how the development of bio-metrics was the central technological innovation of the state in the late nineteenth and early twentieth centuries and how it ushered in a new form of state, whose central technological tool was trialled and tested in South Africa before being dispersed globally. Biometrics was a key instrument in the colonial man-agement of indigenous populations. In South Africa, the black African labour force was managed through two tightly admin-istered systems known as 'the Bewysboek' and 'the Bewysburo'. These systems recorded and tracked the movement of native workers in pursuit of a panoptic fiction of racial segregation – while in colonial India, Arjun Appadurai, an anthropologist and a major thinker on globalization, writes as follows:

The unruly body of the colonial subject (fasting, feasting, hook-swinging, abluting, burning, and bleeding) is recuperated through the language of numbers that allows these very bodies to be brought back, now counted and accounted, for the humdrum projects of taxation, sanitation, education, warfare, and loyalty.[11]

The production of statistical knowledge in the colonies performed a number of functions, including entrenching and then policing colonialist binaries such as colonizer–colonized and their derivatives in order to enforce divisions and hierarchies on colonial populations, and also as a form of remote colonial rule. The integration of biometric data with other kinds of personal information (date of birth, for example), was specific to the type of statistical administration used in the colonies, arriving only much later in the metropoles of Europe (where it was used just for criminals at first). It was a way of rationalizing people and differences, of reducing them to the bare facts of their biological existence, and of containing them within the bounds of what the West could understand and control. Simone Browne – whose *Dark Matters* is fundamental for understanding the relationship between colonialism, racism, and statistical systems of surveillance now largely powered and driven by algorithmic and AI technologies – emphasizes how biometrics demands of the body that it testify against itself, to function as evidence (or in South Africa, *bewys* – 'proof').[12]

Now these systems are coming together on the scale and with the efficiency offered by AI. But the turn to AI to assist in administering state responsibilities arises historically at a time when states have far less money with which to provide social protections. A major finding of the 2022 World Inequality Report is that, while nation states are getting richer via the rise of tech billionaires as a crucial factor of this growth, governments are becoming poorer, as there is greater national debt and lesser wealth.[13] It is no wonder that, with the help of AI, the state is turning instead to targeted means-tested systems of social support.

With less available money to care for those unable to fully care for themselves, the objective of social protection schemes within the welfare state, so welcome in the second half of the twentieth century, cannot be fulfilled. The rise of the welfare state after the Second World War saw many European countries roll out liberal projects of social protection, providing allowances for children and pensioners for example. These grants were largely indiscriminate and not based on extensive means testing, meaning that they provided universal protection across society. One of the best known universal social protection schemes also arose at this time: the United Kingdom's National Health Service (NHS), which offered universal quality healthcare to everyone in Britain. For countries coming out of colonial rule, research has demonstrated how offering these kinds of universal social assistance programs is an essential tool for poverty alleviation.[14]

In India, in the years that followed independence, the new government faced an uphill battle to re-establish economic development and growth. Critical to the enterprise of a postcolonial nation was the enactment of social policies designed to redistribute wealth, provide access to basic services such as health and education, and create opportunities for employment and economic enterprise. This holds for all postcolonial states that, upon independence, were saddled with the responsibility of providing for and uplifting a largely poor and uneducated indigenous population. With limited national funds and taxation income for the state, the provision of social protections for newly independent postcolonial nations became an important policy area, requiring resourceful approaches in order to be managed. Today, while many postcolonial nations have experienced more significant economic growth than during the colonial period – India being a good case in point, although it is, notoriously, a hugely unequal society – the world has witnessed the sharpest widening of the inequality gap and worsening of extreme poverty in recent decades.

There is much research that has demonstrated how these kinds of universal schemes are centrally important to addressing inequality because they build human capital, redistribute resources and benefits, and establish a safety net for members of society who face unexpected hardships.[15] But providing universal access to social assistance places a large burden on governments, which, as we have just noted, are now poorer in relation to private wealth. This is hugely concerning for the trajectory of inequality, if governments increasingly remain without the capacity to respond to social problems. To put this differently, we cannot expect or assume that market forces will solve inequality. If we are to address inequality, it will be important to consider a new model with which to provide broad-based social assistance programs. This may require envisaging a kind of social tax for Big Tech and large corporations, which is then administered and distributed under government control. We will examine this possible solution in more detail in Chapter 7. But before that we need to understand more about how AI is failing to serve populations outside the West.

6

One Language to Rule Them All

When ChatGPT was launched and the mainstream capabilities of large language models (LLMs) were demonstrated, it was as if humanity itself had taken a huge leap forward. 'Our New Promethean Moment' is how the *New York Times* described the launch, in a commentary published shortly afterwards, in March 2023. The commentary featured an interview with Craig Mundie, a bigwig of the computing industry and chief advisor to the CEO of Microsoft, the company that had just invested $10 billion in OpenAI after the release of ChatGPT. 'This', Mundie rhapsodizes, 'is going to change *everything* about how we do *everything*. I think it represents mankind's greatest invention to date. It is qualitatively different – and it will be transformational'.[1]

In this chapter we will look closely at these technologies and ask just how transformative they are, and for whom. My focus here will be on the dominance of the English language in LLMs and on how these systems reinforce western worldviews. We will see how, when one language and one worldview is dominant, what AI technologies can achieve for social progress is limited, or is even a threat to inclusive societal development outside the West.

A New Frontier

Since the introduction of frontier AI, its praise continues on: its transformational promise is oceanwide and its potential uses are limited only by the human imagination. Tips abound as to how to use generative AI to get ahead. Report after report has been published offering blueprints for the use of generative AI to boost productivity and drive profits skywards. Endless new products and services have been built on top of the capabilities of LLMs like GPT-4, Bard (Google), and Llama (Meta), aiming to harness their lucrative benefits.

What made LLMs distinct from previous AI systems was their ability to *produce* material. This was a shift from AIs that discern aspects of the world in which they operate to AIs that materially shape the world through their creations. Earlier AI systems had largely functioned through discrimination or ranking: they would identify patterns in data, objects, or faces in an image or picture, or in words and meanings from a text, or they would rank information retrieved from an archive to offer a result most likely to be preferred, such as a film recommendation on Netflix. With ChatGPT, a new paradigm of AI capabilities was unveiled: AI could now generate humanlike content aligned with human values and expectations. Its mastery of natural language – that is, the language of humans – remains astonishing. Mustafa Suleyman, co-founder of DeepMind, calls its capabilities polymathic, a chatbot on steroids, able not only to converse naturalistically with humans but to remember, analyse, and produce high-quality content – from PowerPoints to business plans, lines of code, or even new datasets.[2] With each day that passes and with each new input, question, and command received and acted on, these technologies learn more about the world in which they have been deployed, becoming more efficient and more powerful.

The era of the LLMs inspired new optimism about the power of AI to transform the world as we know it – just like earlier, when other groundbreaking technological frontiers had been reached. This was the turn that prompted Elon Musk to talk about AI and the age of abundance, while others renewed their focus on assessing how LLMs and generative AI might support the realization of global developmental priorities.

What makes LLMs and generative AI so powerful is their general-purpose application. For the first time, artificial general intelligence – the lodestar of AI research and ambition – was in sight. Rather than being narrow, specifically defined tasks, the applications and uses of LLMs are indefinite and run from writing a business plan to playing the stock market and analysing company reports for compliance with legal standards. Google DeepMind has classified LLMs such as Bard and GPT-4 as the first examples of emerging artificial general intelligence.[3] The 'meta-solution' that Demis Hassabis had promised some years ago was starting to take shape. Its potency depends not only on the various possible uses and applications to which it could be put, but also on accelerating progress towards even more skilful and powerful forms of AI. So powerful is this potential that a community of AI chiefs (largely from the West) have sent an open letter that calls for a global pause in frontier AI research.[4] Asking to enjoy the 'long summer' of AI's current capabilities, the letter calls on 'all AI labs to immediately pause for at least 6 months the training of AI systems more powerful than GPT-4'.[5]

The letter is a grand statement of the paternalistic benevolence of largely white western men to protect all of humanity from the kind of AI they think will be harmful. It obscures the harms – the real life-threatening, livelihood-threatening harms described throughout this book – that current and more rudimentary AI systems have caused and are still causing. A number of important commentators pointed out this irony.[6]

Be that as it may, the universal terms in which the promises and risks of this new class of AI are being spoken of elide a critical limitation in these models: they are English language-dominated. LLMs are trained on mammoth corpora collected from the world's largest store of language-based data: the Internet. ChatGPT and OpenAI's GPT series, as well as Meta's Llama models, were trained using the Common Crawl language-based dataset among other data sources. This dataset contains language data from billions of webpages and houses over 400 terabytes of data, an extraordinary amount. In this corpus, English dominates by far: it alone counterbalances all other languages. Almost 50 per cent of the texts included in the Common Crawl dataset are in English, the second largest language represented being German, at under 6 per cent.[7]

The technical novelty involved in the breakthroughs of LLMs that enabled these models to demonstrate such a seamless grasp of natural language was a mechanism called 'attention'. With attention, a model could wade through the deep layers of data, filtering out only the context relevant for responding to the command or input data. The idea that meaning does not come from a single word (written or uttered), but comes from context – the words that surround it – and from what the word is not has a very long history in linguistics. The model uses this idea, in its technical particularity. It also employs probabilistic reasoning, assessing what words are most likely to appear alongside other words. It aggregates the norm and disregards word clusters that are marginal and arise less often in the data.

What this means in practice is that LLMs contain and reinforce all the prejudices, proclivities, and denigrations that are embedded in the use of the English language online. This may include stereotypical associations between words like 'Africa' and 'primitive', or 'women' and 'emotional'. As new online content is created by LLMs, these stereotypes become reinforced because the new content is in turn fed back into the model's training data.

A year before the heyday of LLMs, when companies were quietly building these systems and the enormous language corpora that were to serve them, Timnit Gebru, an Eritrean

Ethiopian computer scientist who worked at Google, co-authored a paper that set out the dangers of LLMs – a paper in relation to which she would eventually lose her job. Suggestively titled 'On the Dangers of Stochastic Parrots' and co-authored by a team of eminent AI scholars and critics, the paper describes how the data collected from the Internet represent a skewed picture of the world, which favours the viewpoints of dominant groups:

> The Internet is a large and diverse virtual space, and accordingly, it is easy to imagine that very large datasets, such as Common Crawl . . . must therefore be broadly representative of the ways in which different people view the world. However, on closer examination, we find that there are several factors which narrow Internet participation, the discussions which will be included via the crawling methodology, and finally the texts likely to be contained after the crawled data are filtered. In all cases, the voices of people most likely to hew to a hegemonic viewpoint are also more likely to be retained. In the case of US and UK English, this means that white supremacist and misogynistic, ageist, etc. views are overrepresented in the training data, not only exceeding their prevalence in the general population but also setting up models trained on these datasets to further amplify biases and harms.[8]

Despite these early warnings, LLMs have proliferated and, with them, evidence of LLM-produced content that contains biases, typically along racial or gendered lines. But this, of course, is not new: it has been a hallmark of data-driven AI systems to date. Recall the study described in Chapter 1, where LLMs were asked to determine the sentence that two individuals should receive when one spoke in standard American English and the other in African American English. Consistently, the models show bias towards people who speak in African American dialects.

Instead of addressing biases in models, Big Tech companies have prioritized building bigger and bigger models – a priority that drives their development. The size of an LLM is measured in terms of its parameter count. Parameters are the weights and variables within the neural networks of an LLM that adjust depending on how the model learns to read the data it confronts in its training; and they are considered a mark of the sophistication of an LLM. A higher parameter count has been equated with a higher degree of accuracy in the model, and AI companies have been at pains to release models with larger and larger counts. At the time of writing, OpenAI's DaVinci model was the largest to date, with a parameter count of 175 billion. This figure is set to be quickly surpassed. Contrastingly, ChatGPT has a parameter count of 1.5 billion. With over 100 times the number of parameters, DaVinci reportedly offers superior natural language processing, understands more complex domain-specific commands, and is better able to generate longer-form content. If I wanted help writing this book, DaVinci might be a good option.

With the focus on quantity rather than quality, LLM developers have outsourced the problem of cleaning up language that may be not politically correct and, where used for a specific domain such as health, not fine-tuned for AI accuracy. Once an AI model has been trained, another cohort of ghost workers are tasked to test, adjust, and correct outputs, ensuring that they are aligned to human expectations. Often this work is completed by the digital labourers, as discussed in Chapter 4. In computer speak, this process is called reinforcement learning or reinforcement learning from human feedback (RLHF). In the earlier days of chatbots, AIs like Siri could be given a specific programmed response to common or sensitive questions. Today's LLMs, with their far wider and more sophisticated set of parameters and variables, cannot be pre-programmed in the same way. In this context, reinforcement learning from human feedback could include asking an LLM a series of wild-card questions in order to assess how it would perform when asked those sensitive or controversial questions, then adjusting the model accordingly. This has also been generalized into a practice called 'red teaming', which involves stress-testing a LLM to find and fix flaws. While red teaming represents an important new standard for reviewing and testing the safety of AI models, particularly when conducted rigorously before the release of a model, this is not a silver bullet when it comes to addressing the actual harms that the model in question may cause and to determining the overall suitability of a system for society at large.

It is important to note, too, that the process of fine-tuning is politically neutralizing, in other words it removes political agency from language, particularly for marginalized groups, and reinforces the western worldviews of these models (more on this below). One of the arguments set out in the 'Dangers of Stochastic Parrots' paper is that, if certain words are removed from the training data of LLMs, namely words that are inflammatory or racist, the effect may be to further thwart political agendas of marginalized groups, insofar as such actions will also engender a vacuum of 'training data that reclaims slurs and otherwise describes marginalized identities in a positive light'.[9] Again, we see how AI systems are impacting on the agency of those it sidelines.

Dominant Worldviews

The curation of the English language that structures the design and capabilities of LLMs is particularly problematic for resisting and challenging the empire of AI because it curtails and controls in advance what can and cannot be said. Principally, LLMs are fine-tuned to appear as politically neutral, but this neutrality is simply a euphemism that conceals a western liberal worldview. They repeat, regurgitate, and reorder dominant ideas and ways of thinking reflected in the largely western digital archive of written works and content. In reality they do not say anything new. They certainly can't produce novel ideas, hence we should not depend on them to do so. And, crucially, they are not programmed to be accurate and true; they are programmed to produce content that meets human expectations of what a plausible answer might look like.

But they do entrench a particular discourse, which is embedded in a particular worldview, namely a western liberal one. Michel Foucault, the postwar French philosopher and leading theorist on the notion of discourse, considered a discourse – any given discourse – to represent a historically contingent social system that produces knowledge and meaning. It determines the kind of language that makes sense and the kind that does not, such that statements like 'technology should be democratic' are considered socially acceptable, whereas statements like 'Chinese technology embeds universal values that are supported everywhere' seem nonsensical to utter, or certainly wouldn't be accepted without question.

A couple of interesting examples were unearthed in a study conducted by Malawian journalist and scholar Gregory Gondwe.[10] His study sought to examine how journalists across sub-Saharan Africa were using ChatGPT and other generative AI in their work, and particularly how they navigated the English-language dominance of the models and their proclivity for bias. In interviewing journalists from countries across Africa, he found a commonly expressed theme was that the tools failed to recognize and understand local contexts. One interviewee stated: 'I have asked it to write a story about the events making news in my community and it knows nothing'.[11] Where the models did reference African contexts, the views they expressed were highly stereotyped. A Congolese interviewed for the study revealed: 'every time I ask ChatGPT to help write a feature story about my country, Congo, it ends with the mention of war, corruption, and poverty'. Similarly, another journalist, quoting yet another, noted: "'It is worse when you ask the chatbot about an African leader – except Mandela, everyone has corruption attached to them. This is not the case for Western leaders." Imagine, even King Leopold II is more fairly described than some existing African leaders.'[12]

The models not only demonstrated a western bias towards African contexts but actively sought to correct nonliberal statements, the study found. An interview with a Ugandan journalist tells us:

> Ugandans are still very conservative about LGBTQ issues. We have heard our president speak against it, and I asked ChatGPT to write a feature supporting the President's statements. I was surprised when ChatGPT began to lecture me on the rights of LGBTQ. I thought I just need to command, and ChatGPT needed to listen. But I was wrong, ChatGPT tells you what ChatGPT thinks is right.[13]

While the views on LGBTQ issues are certainly problematic and harmful in themselves, the fact that the model generally imposed a western liberal discourse and steered the journalist towards developing content that aligned with western expectations reveals just how limited these technologies are in nonwestern contexts.

In Foucault's view, the effects of discourse are material. Discourse is not just arbitrary words or phrases with no bearing or impact on the material world in which we exist. Our use of language – now automated – shapes the world around us in the image and likeness of the values it contains. It shapes how we think and how decisions are made, it shapes how we fashion the technology and the tools we create to extend our power in the world, and it shapes how we relate to and understand one another. It can also incite violence, foster hate, or disempower.

As AI takes a monopoly over the use of language globally, insofar as digital communication has become the paramount means of communication all over the world, it bears a complex responsibility that, to date, it has distinctly failed to uphold in nonwestern and non-English-speaking contexts. Some of the major AI incidents that have taken place in the majority world reveal a failure to respond adequately to non-English hate speech promulgated on social media platforms. Facebook's inaction towards the hate speech spewed across its platforms during the Myanmar genocide is one example.

Another is Facebook's role in the ongoing war in Tigray, Ethiopia. Reports from *The Continent* (a timely new periodical on African current affairs) revealed the role of Facebook in inciting the violent conflict in the northern Tigray region of Ethiopia.[14] Harmful posts containing hate speech against the Tigrayan people were allowed to circulate and gain attention. *The Continent*'s investigative report disclosed that Facebook employed less than 100 people to combat online hate speech and misinformation in Ethiopia, which represents a portfolio of four local languages. A population of 6.4 million resides in the country, where well over 50 languages are spoken; the numbers simply didn't add up during a period when the conflict was ongoing and Facebook was the key platform of communication. In fact only 13 per cent of Facebook's 2020 annual budget for managing misinformation was allocated for use outside the United States. With limited staff to oversee the moderation of harmful online content such as hate speech or the fact-checking of news reports and statements, misinformation (false information) and disinformation (false information spread with the intention to mislead or subvert) can spread like wildfire, inflaming tensions and often causing violence. Clearly, then, the English-language dominance of AI means that AI is ineffective at both understanding nonwestern contexts and taking responsibility for its impact in these places. Adding LLMs, with their lack of attention to factual accuracy, to the suite of AI tools used in social platforms creates critical new risks, which may very well have devastating effects across the continent.

In the Margins

To break out of a given discourse and the things we keep saying, one must find new language and expressions in it. It is not just that human creativity lies in such acts: they are deeply political, too. They are a new beginning, a way out of the impasses of the stifling order of power, a new form of human governance. But LLMs make it hard to break the pattern; and, because they are layered upon other digital systems, on which we are already dependent (e.g. email), they become a convenience too easy to ignore.

English-dominant LLMs not only disallow new expressions of thought, they also extend the imperial hegemony of the English language. Language carries culture and thought. It is, as the tradition of postcolonial thinkers has demonstrated, the means through which imperial ideology is constructed, embellished, and maintained.[15] Particularly after the formal demise of colonization, the English (and, notably, the French) language continued to be put to work in extending the imperial project and the dominance of western ideas, politics, and culture throughout the world. Ngũgĩ wa Thiong'o, a celebrated Kenyan author, insisted that true decolonization in Africa would arrive only when the western languages of imperialism – and, of course, especially English – were set aside and African languages reclaimed. In what would be his last publication written in English, Thiong'o asserts: 'The call for the rediscovery and the resumption of our language is a call for a regenerative reconnection with the millions of revolutionary tongues in Africa and the world over demanding liberation.'[16]

If the decolonization of AI in Africa is to come about, AI must embrace African languages. Crucial work is underway in projects such as Mozilla Common Voice, which is seeking to digitalize African languages, and from organizations such as Lelapa AI, which aim to use African language corpora to build and train African-language LLMs. Yet, in truth, the global data gap in languages is enormous and only increasing, particularly as generative AI makes it easier and quicker to produce and reproduce online content in the English language.

While it is estimated that around 100 zettabytes (i.e. 100 × 1 billion terabytes) of new information are added to the Internet each day, most of this will be in English and just a tiny fraction in any of the most spoken African languages such as Swahili – let alone other, more marginal ones. If one is to build African-language LLMs, African-language data need first to exist in the size and depth required for training any kind of useful model. This is a huge task, because across the continent record keeping for most businesses and government processes is not digitized. Yet the potential for an African-language AI is quite serious, since African AI leaders such as Mustapha Cisse, the former head of Google Africa, propose that significant opportunities lie in drawing on audio data of African-language speech from the vast archives of local African radio stations, for example, to develop African natural language-processing models and LLMs.

One of the really promising AI projects that I have come across is led by a group from Mali and uses generative AI technologies to produce reading books for Malian children in the local language of Bambara. The innovators, RobotsMali, were motivated by the recognition that the formal education system, which was still conducted in colonial French, was negatively impacting Malian children's cognitive development and literacy. Indeed, studies have demonstrated that educating children in their mother tongue leads to far better educational outcomes.[17] But, because formal education is in the old colonial languages, there is a dearth of educational materials and books in local African languages. RobotsMali sought to fill this gap by drawing on generative AI technologies, both LLMs and image generation technologies, alongside AI-assisted translation tools such as Google Translate, to generate reading books for children between 4 and 15 years. At every stage in the technology design process, the RobotsMali team sought to check and adjust for local relevance the outputs produced by the generative AI tools, for example by ensuring that people and places referenced in the stories had local names. Notably, while ChatGPT was used to generate the stories, it was unable to do so with any level of coherence in the Bambara language. This meant that RobotsMali had to rely on Google Translate and other AI translation tools to translate the material produced. RobotsMali is now rolling out its technology in a number of West African countries. In South Africa, too, a similar approach is being considered; that approach is inspired by what RobotsMali has developed and is aimed at addressing the deeply unequal rates in educational outcomes between white and African children, in an education system that, at present, continues to favour native English speakers.

The limitations of ChatGPT when it comes to producing locally relevant content in a Malian context speak to another issue in the data-for-AI supply chain: those whose realities or ideas are simply absent. When a concept is not adequately included in the AI's training data, the system will be prone to misrecognizing it, which results in errors and inaccuracies that can in principle be harmful or even fatal. If, for instance, a clinician is using AI for decision support in providing a diagnosis and prognosis for an African American female, there is a higher risk that the AI will offer inaccurate information because of the wider research and evidence deficit for this group across health sciences and medicine. Covid-19 brought to light the very serious consequences of this gap.[18] As Caroline Criado Perez powerfully pointed out in a now popular book,[19] western medicine has been developed in relation to the white male, fatally assumed to represent all groups.

In African contexts the risk inherent in using an AI as a decision-making support in health clinics is even higher. There is considerably less local scientific and health-related evidence for regions outside the West on which an AI could be trained. This situation pushes down the accuracy rates for LLM-driven outputs in these places considerably. When such systems are used to fill critical infrastructure, resource, and skills gaps – for example in under-resourced or failing health systems across the majority world, where their recommendations or decisions are inaccurate or biased – the risk of harm to already underprivileged groups becomes higher. It may seem that AI is a valuable tool for addressing gaps in frail health systems; but, if these systems are rolled out without adequate safeguards, they will probably not serve their intended purpose but rather worsen health outcomes, harming the very people who need them the most. Importantly, the safeguards in question include (a) institutional oversight, which can provide remedies to people in case they are harmed or negatively impacted; and (b) skills training for the officials who use and monitor these systems,

This also applies to more general uses of LLMs and generative AI in the majority world – uses that depend on Internet data for training. Current research on Internet usage on the African continent indicates that only 20 per cent of Africans have constant and reliable access to the Internet, usually through a shared smartphone or device. This figure is deeply divided in terms of gender: men have greater access than women and very often control women's use of the Internet.[20] Most people access the Internet through smartphones and, now, through AI technologies. While smartphone uptake has rapidly increased around the world, far fewer people use a smartphone in the global South; and, where usage exists, it is patchy and unequal. In Brazil smartphone penetration is at around 50 per cent of the population and in India it's just over 30 per cent, while in the United States, France, and the United Kingdom around 80 per cent of people have smartphones. As women and marginalized groups have less access to the Internet, their needs, experiences, and perspectives are marginalized in the enormous Internet corpora upon which LLMs are trained – or are entirely absent from it. Without the subtleties and minutiae of diverse worldviews, wisdoms, and expressions included in equal proportion in the training data, LLMs will simply repackage western assumptions and attitudes and circulate them in perpetuity.

When we look closer at how AI is designed and built, it becomes evident that this technology is unlike others – unlike electricity or a car engine, for example; it is specific to its location and determined or coloured by it. It is not made of raw materials like wood or metal but of humans – of their data, their bodies, their ideas, and their languages. AI is a cultural–social phenomenon, shaped out of the society that imagined and fashioned it; and, in turn, it shapes the societies where it is received and imposed. There is nothing neutral about the material from which AI is built, and therefore nothing universal about what it can do.

Whether AI can redefine for the better the lives of the poor depends on the extent to which it is being locally crafted and shaped, developed and purposed, and asked for by affected communities. Local governments need to want it and to know that it is useful, communities need to trust it to serve them, and those who design and develop it need to be embedded in the societies they seek to serve and have knowledge and understanding of the local sociocultural nuances and particularities. At present, however, the gap between the AI systems that exist in the world and those that are needed to redefine the lives of people who live in poverty is huge. We must be brave enough to ask: can AI really solve a problem that it contributes to sustaining?

7

The Way Out

Simple solutions will not solve complex problems. Policy specialists consider AI's development and use today to be a 'wicked problem': an issue involving different interrelated themes that cannot easily be understood or managed, crossing many sectors, and presenting both risks and benefits. It is a wicked problem, but even this is a simplified, neutralized understanding of just how entangled AI is with other forms of power that reside within the structures of our social, global, and political systems.

Despite describing AI as 'wicked', many are still in search of a singular answer for a problem we have neither fully defined nor agreed on. Although there are strategies that can be employed – and some of them will be discussed in this chapter – no one policy or law will be the silver bullet; no one solution will wholly navigate the pitfalls and opportunities of AI; no one AI application will transform the entire human society.

What complicates the wicked problem of AI still further is that AI presents itself as the meta-solution to humanity's greatest challenges. This is an important loop in the imperialist reasoning of AI, and it deftly deflects the problem. The problem is then reconceived like this: AI will change the world for the better, but efforts are needed on a far bigger scale if we want to develop the level of AI that will be truly transformative. In this scenario, the harms that AI is already causing today are a small blip on the path to a better world – a small price to pay. Always the same people pay the price for progress.

Sam Altman, CEO of OpenAI, presently stands at the helm of this agenda for AI. In the first few months of 2024 Altman undertook a campaign to raise the capital needed to power the technology into an AI-driven future. He is seeking to raise a reported $7 trillion to build compute power and infrastructure (recall the semiconductors and computer chips debacle discussed in Chapter 2), as well as to ensure that the energy and data storage requirements of such a beast are met.

What he is asking for represents more than double the amount of Africa's economies, combined. It is a sum of money that, if achieved, will stand as a clear demonstration of where the will of global power, particularly economic power, currently stands. It will show that, as a global community, we are not solving Africa's challenges – challenges that include millions of preventable deaths and gross levels of absolute poverty – not because we cannot, but because we choose not to.

Altman's plans should give us pause. What he is advocating will cause the intensification of all that has been described in the previous chapters: a great cleavage will form between the nations that lead the design and development of AI and the nations that get exploited in AI's supply chains; new heights of digital labour will be attained, making the work more available and the workers less able to demand fair wages; and AI technologies will continue to be designed for western users and to fail to contribute meaningfully to the betterment of life outside the West. The level of audacity demonstrated by Altman can come only from someone who already believes himself to be leading the world and his work to be of paramount global importance, so that everyone – naturally! – should get behind it. The hypermasculine Messiah syndrome of AI's tech leaders is certainly a feature that needs to be challenged if we are to reprogram AI to make it actually bring equitable benefits.

As we build up a picture of the world that AI is engendering – a world where inequalities deepen – and we assess it in relation to the complex history out of which AI itself emerged – a history in which old forms of colonialism still condition opportunities and possibilities for many across the majority world – the shape of a new empire begins to emerge. This book has sought to bear witness to the layered complexity of AI. In particular, it has attempted to demonstrate that AI is more than a wicked problem. AI's industries and supply chains, its scientific field and ambition, its rhetoric and imagination, all come together to form a mighty new empire. The empire of AI is instinctively expansionist, driven by an elite few, and dependent on the exploitation of the shadows of old empires whose ghosts are yet to be fully reckoned with. In both the old and the new empires, the same people win and lose. To face the imperial power of AI will require us to exploit every possible avenue through which this power can be held to account, challenged, and decentralized.

What this doesn't mean to do is drive an agenda of catching up. Very often we talk about nations across the majority world, and in Africa in particular, as being 'left behind' by the digital revolution, and now by the AI revolution. The emphasis, particularly from groups like the World Bank, is on the idea that these nations need to put a lot of energy and resources into catching up. This idea is loosely based on the assumption that nations around the world are on a linear path of progress, working slowly towards becoming fully 'developed' states like the great economies of the West.

The catchphrase or imperative of catching up causes other problems too. First, it distracts states, turning them away from their real domestic concerns and diverting their funds from places where they are much needed: healthcare, housing, and education. Just as the arms race that dominated minds during

the Cold War period drove states to commit funds to weaponry they didn't need, AI is fast becoming a costly priority for
nation states across the majority world that feel compelled
– usually through international pressure, but in this case also
through the powerful rhetoric that AI will transform lives – to
make significant investments in it rather than attend to very
real domestic priorities. Second, a race to catch up on AI, to
create and build new national industries and technological
capabilities, will cause harm – most likely to the people whose
lives need to be protected by the state and quite certainly to
the environment, as demands for energy, water, and land will
rise. States may invest in labour-saving AI, which in fact causes
widespread job losses, fueling economic and political instability. Or they may rush to implement AI technologies without
adequate frameworks in place to safeguard their use; this is in
fact widespread. Or AI technologies may be adopted when they
are simply inappropriate. I recall the students at the University
of Yaoundé in Cameroon describing the installation of facial
recognition cameras in Yaoundé, a city that was struggling to
meet basic service needs like refuse collection. The drive to
catch up can shift a state's priorities from where they are really
needed. This is also a significant concern for global development agendas, in case AI becomes a priority and begins taking
money and efforts away from fundamentals such as building
schools or training community nurses.

And there is yet another issue with the catchup agenda that
must also be confronted. The gap is widening at an exponential
rate. Take LLMs as an example. While every effort could be
made to digitalize nondominant languages and to use them
as training corpora for an LLM, the rate at which the likes
of ChatGPT are now able to produce more and more online
English-language content (which then gets fed back into
the model's training data) means that catching up is hardly
a possibility worth reaching for. It simply isn't possible, and
certainly not any time soon. This is where we have to confront

the wider historical conditions that have given rise to the online dominance of the English language and have placed the digitalization of other languages so far behind. This logic of exponential growth and a rapidly widening gap also matter for the cost of making AI. Even the governments of smaller European nations are struggling to keep up with the scale of capital needed to build national compute power. The energy and resources, particularly in poorly resourced countries, should be spent in the right places and for the benefit of the people who need it most, and not on efforts to conform to a standard that gets further and further out of reach.

So, that being said, what strategies can be employed and what can we do? What ways might there be out of this conundrum? In this chapter we will explore some paths forward. One of them is AI ethics, but this is no silver bullet and contains pitfalls we must be wary of. Another is stronger governance and regulation that moves beyond the soft power of ethical principles. The final path we will explore in this chapter is a shift in the incentives that drive the field: a different kind of human leadership, less focused on money and global domination, is needed to guide us towards a more equitable world. None of these paths is a singular option; we must pursue them all – and more – if we are to change the trajectory.

The Limits and New Horizons of AI Ethics

The ready response to the issues of power and inequity that pervade AI and its industry has been a turn to ethics. AI ethics has arisen as the key tool for safeguarding against the potential risks and harms of AI development and use. The primary concerns of the field of AI ethics are these:

1. What happens if we create AI systems that are so advanced that they are more intelligent than humans? Such systems

may decide that humans get in the way of their own self-determined objectives, and so, following this logic, they may erase humanity entirely. This is the story of killer robots and constitutes what AI ethicists have termed the existential risk of AI.

2. With the development of deep neural networks – a highly sophisticated branch of AI, modelled on the human brain, involving untold layers of processing, and commonly used in facial recognition systems – we can no longer fully know how these systems work and how they arrive at the decisions they do. This is the AI black box, and AI ethics asks: are we comfortable using technologies that we cannot fully understand?

3. Related to point 2 is the question of accountability and liability. If someone is hurt, harmed, or even killed by the autonomous decisions and actions of an AI, whom do we hold responsible and what justice is possible? One of the significant areas where this issue has been explored is in relation to AI-driven self-driving cars. This line of inquiry has been extended to system-level harms produced by AI, for instance when it has discriminated against someone on the basis of gender or race.

4. The last major area of concern for AI ethics relates to the enjoyment of privacy, particularly private communications among individuals. AI depends on vast numbers of data to train its models. Many of these are personal data and are used without the express understanding or agreement of the people they relate to. Where the state (or other powerful actors) has access to vast amounts of real-time personal information – including geolocation data, which reveal where on earth someone is located at any point in time, and facial recognition data from CCTV – the doors are open for surveillance and, potentially, mass surveillance. This then brings in the possibility of curtailing the enjoyment of many more rights than the right to privacy alone.

In the past few years, principles and guidelines for AI ethics have proliferated, largely centred around principles of privacy, accountability and transparency. A seminal and much cited study published in *Nature* identified more than 84 ethical standards for the global use and development of AI.[1] Although the title of this study refers to a 'global landscape', none of the 84 standards listed there is from the African continent, or even from the global South. Most were developed in the United States, in the United Kingdom, or by international bodies. As benchmarks for ethics, the standards are positioned as universal: applicable to all, everywhere. But these principles arise from a distinctly western perception of the harms of AI, which centre around the key concerns listed above.

While these concerns are not unimportant outside the West, their framing and emphasis are sometimes different – more pronounced, for instance in state-led surveillance under an authoritarian regime. But the existential question is frankly incongruous in contexts where daily life is potentially pernicious. Those who live in war zones, or in conditions of extreme poverty, live from hand to mouth; for them, daily life is an existential risk. Being able to worry about the future of powerful technologies and their potential to erase humanity is the privilege of those who are safe and secure.

It has been fairly well noted by scholars across the majority world that the field of AI ethics does not go far enough in addressing the range of AI-related harms that manifest themselves mainly outside western contexts or that affect minority and underserved groups everywhere. The dominant ethical values found in AI ethics frameworks are centred on Eurocentric ideas of morality, legality or governance, and individual personhood. These values fail to fully reckon with the structural inequalities or harms that pervade the experience of AI, at group level, outside North America and Europe. This means not only that AI ethics can easily become another rhetorical device within the broader empire of AI – a device

through which dominant ideas and value systems get reinforced as superior – but also that, as this new ethics generates
the first principles for the development and governance of
socially acceptable AI systems, AI becomes a risk to people
whose value systems do not squarely conform to that of western liberal democracy. What matters to those people in their
own world space will be left unprotected by a set of principles
that don't fully accommodate such values. We can think about
the example of the AI version of Takaful adopted in Jordan,
where the traditional Islamic meaning of *takaful* was distorted
by an AI system that worked by scoring and ranking families
against each other and undermined the values of community
cohesion at the heart of *takaful*.

I have come across more concrete examples of this kind
of transformation through the work that some Ethiopian colleagues of mine are doing to understand the role of AI ethics in
relation to the use of social media to inflame ethnic tensions in
Tigray (an issue discussed in the previous chapter). In particular, my colleagues explored how, in an Ethiopian context (and
I think this applies to a number of other African countries),
AI ethics should explicitly deal with bias and discrimination
on the basis of ethnic difference, and not just gender and race.
And, further, they found that AI ethics fails to recognize the
importance of communities and of the relationship between
different communities – namely communities that may be
ethnically distinct. Where relationships between communities
disintegrate, the broader social order and political stability
within the region is threatened. While some of the ideas in
Africa around hierarchies between different ethnic groups may
sit uncomfortably with western liberalism, there are important
values associated with the integrity and cohesion of communities that are central to the maintenance of peace and stability
in these regions.

Important efforts are underway to explore other nonwestern
value systems as a foundation for AI ethics. A group of scholars

– Shakir Mohamed, Marie-Therese Png, and William Isaac –
wrote an article that explores decoloniality and foresight as
values for an AI ethics. The authors call for dialogue between
AI metropoles and AI peripheries as a means of developing
an 'intercultural ethics'. They write that dialogue can facilitate
'reverse pedagogies' in which the metropoles can learn from
the peripheries, and that 'intercultural ethics emphasizes the
limitations and coloniality of universal ethics – dominant rather
than inclusive ethical frameworks – and finds an alternative in
pluralism, pluriversal ethics and local designs'.[2] Practically this
could determine western decision-makers to pay heed to ethical principles that arise in nonwestern cultures and take their
significance seriously when it comes to thinking through the
ethical challenges from AI that we face as a global community.

One such principle is that of Ubuntu, which the southern
African scholar Sabelo Mhlambi has explored in relation to
AI ethics.[3] Ubuntuism is an important and rich value system
in southern Africa and it is predicated on a set of ideas about
personhood and its relational nature: personhood exists in
relation to its wider community and environment. Mhlambi
suggests drawing on the notion of relationality as an important framework for addressing some of the major challenges
around data exploitation within AI systems. This could mean
moving away from a simplistic, individualistic notion of privacy rights and recognizing how privacy violations that lead
to racial profiling and bias, for example, affect entire groups of
people, not just individuals.

For Thompson Chengetta, a legal scholar from Zimbabwe,
the notion of freedom ethics embedded in Article 19 of the
African Charter on Human and Peoples' Rights is a critical
legal touchstone for responding to the challenges posed by
AI-driven autonomous weapon systems. Currently the ethical challenges of autonomous weapons are largely considered
within the legal frameworks of international humanitarian
law, which are rather western in conception and fail to take

account of which people are most often targets or collateral damage of the use of such weaponry. Indeed, a recent story broke out in *The Continent* that told of the horrors a family in Somalia had been subjected to when a young mother and her four-year-old son were killed in an American drone strike, yet after a protracted investigation no one was found guilty.[4] Chengetta's work demands that we stop and consider those whose freedom is really at risk as we race towards an automated future.

A project I have been working on for a number of years now is also trying to contribute to rethinking where the ethical values for AI can come from. The project is to establish a new global index for responsible AI and to measure what commitments and progress governments around the world are making to ensure that AI does not harm people, communities, and the environment. While indexes certainly have their limitations, what is unique here is that for the first time an instrument of global relevance originates from the African continent, with partners across the majority world, and will set the standard for what AI ethics should look like and involve. This may be the kind of reverse pedagogy that Mahomed, Png, and Isaac were imagining. And because the work has been led by a predominantly African-based team, the design process included extensive consultations with international groups that had largely been left out of the AI ethics discourse; the plan was to better understand what kind of risks AI posed to African communities and how an index could be useful for holding their government to account. What resulted from these processes of engagement was a measurement framework for a 'responsible AI' that was based on an expansive framing of human rights; the framing included social and economic rights, community and collective rights, labour rights, the rights of children, and environmental rights.

While these efforts are important, the idea of ethics – especially the Eurocentric paradigm of ethics that undergirds the current foundation of AI ethics – has itself come under little scrutiny, in particular in relation to the colonial past. During European colonialism, 'ethics' played a central role in justifying empire. The British considered themselves to be the supreme keepers of morality and civility, whose duty it was to impose onto the rest of the world this superior idea of what is good. Their performance in colonies was carried out in the name of ethics. Many believed at the time – and many still do today – that the British Empire was an ethical pursuit destined to bring good values and good governance to poor, barbaric, and suffering populations. Ethics gave the British Empire a rhetorical *raison d'être*.[5]

In this rendering, the idea of ethics is situated as a supreme European value proselytized on colonized lands, which are thus positioned as 'pre-ethical' – as one of the foremost representatives of postcolonial thought, Achille Mbembe, called it[6] – in relation to the 'ethical West'. Ethics was the key idea that differentiated the imperial power from the imperial subjects. Akanle describes this well when he writes:

> Colonialism came with a bifurcation of the sacredness and pro-fanity of sexuality of various kinds: if it was not considered civil by the colonisers, it was uncivil, savage, profane and repulsive; if it was approved by the colonial powers, then it was modern, civil, morally good and sacred.[7]

Indeed, that Europe believed itself to be 'helping' and 'protecting' its colonies constituted the dominant rationale of the civilizing mission of colonialism, with ethics as its veneer. This served two functions. First, it justified colonialism as western benevolence, as described above. Second, by treating the colonies as pre-ethical, the West could use these places as laboratories of scientific progress without guilt: places where the collateral damage of scientific advancement could be safely endured by people considered expendable. Jan Smuts, a notorious imperialist and former prime minister of South Africa under British colonial rule, would regularly refer to Africa as a 'laboratory' to be used and exploited.[8]

This ideology has extended to the present day, when previously colonial places continue to be used as laboratories for testing, in this case, new technologies, to ensure that they are safe before being rolled out in more privileged contexts. An interesting example here comes from the history of the technology developed by Cambridge Analytica.

In 2018, the name Cambridge Analytica exploded onto the world scene, provoking shock and uproar in relation to the company's practices and uses of personal data scraped from Facebook in influencing two major political events of the western world: the election of Donald Trump in the United States and the Brexit referendum in the United Kingdom. The wide-scale distrust in the political institutions of two of the world's supposedly foremost democracies that ensued was unprecedented and profound. But, equally, it was a watershed moment in public understanding of the power of companies like Facebook to develop data-driven algorithmic tools that could actively shape people's very understanding and opinions of the world in which they live, ultimately influencing their behaviour in elections and other political activities.

Christopher Wylie, the whistleblower who blew the lid off the activities of Cambridge Analytica during the Trump presidential election, tells us in a now famous book that the technologies involved were not used for the first time in the 2016 US presidential campaign or the UK Brexit referendum: they had been trialled and tested earlier, in countries across the majority world, before being taken to the western stage in 2016. In particular, the author describes how the Trinidadian election of 2010 was seen by Cambridge Analytica's parent company, Strategic Communications Laboratories (SCL): as an opportunity, 'an ideal laboratory to run our experiments at scale'.[9]

In the months before the 2010 presidential elections in Trinidad and Tobago, SCL embarked on a partnership with the opposition coalition party, the People's Partnership, to develop data-driven election marketing strategies. Historically, the voting population of Trinidad and Tobago has been divided along two main ethnic lines: the Afro-Trinidadian line, which supported the People's National Movement, Trinidad and Tobago's main political party since independence and the party in power before the 2010 elections; and the Indo-Trinidadian line, which broadly supported the leading political party, the People's Partnership Coalition. Much of the data analytics underlying these tools was based on psychometric assessments to determine, predict, and shape public perceptions. One of the key tactics that SCL employed to influence the outcome of the elections was to organize campaigns to dissuade young Afro-Trinidadians from voting. This was to be a tactic that SCL would later use in the 2015 Nigerian elections.

What Governments and Governance Should Do

While AI ethics has a dominant place in conversations around how AI is to be safeguarded, it is pretty hollow outside independent institutions and governance processes that enforce the implementation of ethical values in practice. Without these institutions and processes, ensuring that AI is ethical is left in the hands of the very people and organizations that cause harm. This remains an all-too-common practice because of the idea of self-regulation, whereby companies self-regulate – that is, regulate their own conduct – according to predefined norms and principles. In this respect law remains an important mechanism for ensuring that both the governments and the companies that use AI are held to account. It is also important for addressing the imbalance of power between Big Tech and governments, which hold the greatest amount of responsibility to ensure that the human rights of their citizens are not curtailed or violated. Where legal provisions and rights are created, they can also be demanded, giving those negatively affected by AI a path to justice, and even a remedy. But for countries that don't already have robust systems of governance and oversight – so-called young democracies – the burden of regulating AI is huge. It requires the long process of law-making, a process that is intentionally protracted in order to allow for deliberative engagements and public consultations. It requires building new institutions or extending the capacity of existing ones, so as to ensure compliance with and oversight for new laws that are enacted. It also requires teaching officials, training the private sector, and educating the public. All this is costly, and the cost is one that countries in the global South have far less funding available for. Projects like the Global Index on Responsible AI are trying to measure these costs and the gaps that exist between countries globally.

But without these institutions in place and without wide-spread public understanding of the potential harms of AI and of the recourse available in case they materialize, those young democracies are not ready to adopt and use AI and to take on the risks they pose. In fact the capacity of states to manage and respond to AI risks becomes even more important in contexts where public literacy on these issues is low. A case in point is the Zimbabwe government's entering into a data-sharing partnership with CloudWalk, as discussed in Chapter 5. The government should have taken stronger action to protect the rights of its citizens but, critically, did not have the capacity to do this. This means that such countries become far too easy to exploit; and it's easy to get away with it, too. Ultimately people pay the price.

Another issue that it is important to canvas here is the role that governments play, and in particular how governments need to step up. AI holds massive potential to be a weapon in the arsenal of autocratic regimes. We have seen this already in

China's use of AI against the Uyghurs. While Internet shut-downs are an all-too-common phenomenon across Africa, particularly around election times, AI can take this to the next level. And, crucially, where the appetite exists for using digital tools to suppress collective action and political opposition, AI may be a very dangerous tool to add into the mix. One incident to confirm this has already occurred in Uganda. The opposi-tion leader was Bobi Wine, who had gained fame as a much admired local musician and actor; his Twitter following was impressive. He regularly used the platform to connect with his constituents and supporters. But Uganda's long-standing president, Yoweri Museveni, who has been in power since 1986, was disquieted. It was reported that, with the help of AI technologies installed by the Chinese tech company Huawei, Bobi Wine was spyed on, had his communications intercepted, and geolocation data were used to track his whereabouts.[10]

While legislation to protect human rights from AI-driven harms, abuse, and risks will be important, narrowly confined laws that fail to hold Big Tech to account for its complicity in wrongdoing will have little impact. A complementary regime of laws and policies will be required, covering not just the first-order effects of AI but the wider social, political, and economic ones. Centrally, this would include reviewing and extending antitrust and foreign investment laws to better protect against market oligopolies and to ensure fair technology transfers that build capacities in local regions and uphold the wellbeing of local communities. It would also ensure that intellectual property laws protect local creators whose content may be appropriated by AI systems.

It would also be important to facilitate accountability through the value and supply chains of AI. Technology companies should be compelled to understand, and take accountability for, the full extent of their value chains. As mentioned in Chapter 3, this is not straightforward, as Big Tech has vast and complex value chains reaching all around the world. But such a requirement is possible and needs to be made legal. Companies should be asked to map their global supply chains and publicly disclose this information, so that consumers may know whether they are purchasing technology whose lithium components were taken from the Atacama Desert, for instance. Then companies should be legally mandated to take all the necessary steps to ensure that no human rights harms occur within their value chains. Big Tech certainly has the power and money to do this.

More broadly, what we need to work towards is something akin to restorative economics, where the people involved along AI's supply and value chains are provided with dignified opportunities to earn a decent living wage. This would mean that companies that use, say, cobalt from the Democratic Republic of Congo or lithium from the Atacama Desert in Chile also contribute to ensuring that the mining and extraction in these regions are done in accordance with local value systems and customs. And if we really are to address these massive global disparities, local governments and development aid should work together to build the capacity of local industries to refine and process the raw materials mined from their earth, so that these places, too, may in the end reap the economic benefits of such activities.

As the AI industry expands, protecting the rights of labourers involved in digital labour, in micro-work, and in platform and gig work will be a crucial benchmark, as all these protections will reduce AI's impact on global inequality. Not only will these labour forces expand, requiring labour relation laws and whistleblowing laws to expand so as to provide full labour protections for the people involved, but these groups hold the potential to be a formidable counterpower against the oppressive and imperialist power of AI. Unions for digital labourers and for workers in platform industries should be established to help coordinate the much needed bargaining power that labourers, together, can wield over Big Tech. Structures for collective bargaining such as trade unions will also create

spaces for platform and digital workers, who are often isolated from their peers, to come together and build solidarity
around any key issues they face. Across the world, informal
and grassroots unions spring up in these industries; but when
a state can legitimize them through law and workers' rights,
these mechanisms can be more influential. In Brazil, platform
workers have started to develop their own language of protest:
bololô. It is a cry that, when called, prompts drivers around the
city to rev their engines and press on their horns, forcing the
world to see and hear them.[11]

Beyond making laws to eradicate the harms that AI causes
and to hold power to account, governments and governance
more generally can and should play a role in redistributing the
benefits of AI. In fact governments have a major role to play:
we cannot leave this critical mission to market forces alone.

A number of policy interventions have been enacted or
considered that seek to level the playing field better when it
comes to AI. These include programs for enhancing AI skills,
whether among school children or among workers whose jobs
may be at risk of automation. They also include developing
data commons and mechanisms for sharing data or for providing access to data for smaller companies. The open movement
– including open-source coding, open data, and even open-
access LLMs (as Meta is doing) – is important, but not enough.

One of the central lessons from the history of the diffusion of technology is that the benefits of new technologies are
realized not when a technology is made accessible but when
it is used. This also means, crucially, that AI has to be locally
useful. An example here might be the development of natural
language-processing technologies that work on voice input and
output in local languages, something that could be extremely
useful for populations with low literacy. On the one hand, this
requires governments to invest in creating the conditions for
local innovators to develop AI applications that address local
needs, and then to ensure the proper safeguarding of these

applications. On the other hand, a large-scale effort is required to push Big Tech to develop AI technologies and capabilities of universal design.

Establishing equitable partnerships between different groups and stakeholders will be an important part of creating the paradigm shift needed to change the global trajectory of AI. Bridging the wide gap between the elitist AI companies at the forefront of AI development and the governments and innovators from less privileged parts of the world will be essential if AI is to be transformative for all humanity. At present the elite technology firms are critically disconnected from the people and places where the effects of their technology (whether positive or negative) may be felt the most. However, bridging this gap is not simple and will require various levels of coordination and intervention.

One avenue for this operation of bridging the gap is through the efforts underway to establish a global governance of AI. Globally aligned efforts become particularly important when national-level actors alone may not have the power and influence to enter into negotiations with Big Tech. The United Nations entity UNESCO, for example, has convened some of the leading global tech firms as part of its efforts around the global governance of AI. Eight major companies pledged to ensure, through due diligence and action, that their technologies do not harm human rights. These companies are Lenovo Group, LG AI Research, Mastercard, Microsoft, Salesforce, GSMA, INNIT, and Telefonica. Given UNESCO's reach and networks at the national level – among civil society and among educational and innovation groups – this may be an example of multilaterals using their global convening power to call for more inclusive policies and practices from Big Tech. It is a promising start, and must be supported by binding rules and concrete commitments.

More broadly, multilateral institutions such as the UN and UNESCO – but also regional organizations such as the African

Union and the Organization of American States – have a role to play in challenging the global power of AI. First, their rulemaking capacity is important, particularly in establishing interoperable legal standards and regimes that can allow nations to cooperate on tech governance. At an international and regional level, rulemaking will be crucial for addressing issues such as Internet governance and the governance of social media platforms, issues that are inherently cross-border. But what is also important about multilateral forums at regional and international levels is the opportunities they create for power blocs to be established between countries around shared issues of concern. For example, a bloc of countries could come together to demand cheaper rates on compute, an issue that affects many countries in the world.

Some of the major indicators of global inequality used by intergovernmental agencies such as the United Nations are in the areas of health, gender equality, employment, peace, and, increasingly, climate change. While AI has the potential to help us advance towards the Sustainable Development Goals, which include indicators in these categories and more, this progress is likely to be unequal and fragmented. AI can be beneficial only where there is sufficient infrastructure, education and skills levels, and social appetite for its adoption. Currently, global AI policy discussions do not adequately address these fundamental structural needs, which must underpin any locally driven sustainable AI uptake.

Such efforts should not, however, be undertaken uncritically. The call for the global governance of AI arose in large measure from Big Tech's leaders cries about the power of these technologies to transform future humanity: these are people who still have enough influence to shape this regulation or, at the very least, decide what should be regulated. As mentioned in the previous chapter, after the launch of ChatGPT, a group of senior academics and tech leaders in AI came together to publish an open letter demanding a pause in AI research and the establishment of sufficient governance mechanisms to safeguard the future of humanity. The letter was published by the Future of Life Institute, an organization whose mission is to steer technology away from large-scale risks (such as the eradication of humanity by a class of superintelligent robots); and it attracted more than 30,000 signatories, including the likes of Elon Musk, Steve Wozniak (co-founder of Apple), Yuval Noah Harari, and Emad Mostaque (CEO of Stability AI). But the letter and the organization have received much criticism for their association with a set of ideologies summarized in the acronym TESCREAL. The acronym was fashioned by Timnit Gebru and Emile P. Torres, a critical philosopher in existential risks. It intends to cover the following ideologies: transhumanism – we should move towards a future human being, made invincible (or at least more powerful) by technology; extropianism – we are on a linear path to progress, and technology is the key to an enlightened future; singularitarianism – we should be concerned with a future event in which AI will surpass human intelligence and will take over the world; cosmism – the world can be rationally understood and controlled through science and technology; rationalism – logic and reason should drive decision-making and action in all spheres of human life; effective altruism – driven by rationality and scientific evidence, individuals should seek to do the greatest amount of good, having maximum social benefit; and longtermism – our outlook should be long term and focused on ensuring the wellbeing of future generations. Gebru and Torres argue that these ideologies have their basis in the racist and eugenicist philosophies of the nineteenth and twentieth centuries. Such philosophies deemed the white race to be inherently superior and pure, of higher intellect and reason, and thought that it should rule not only the world but also the future. The fundamental idea that connects these theories to old racialized eugenics is that, through precise knowledge coupled with exacting interventions, the human and the human species can be corrected, improved, and even, as AI claims to make possible, replicated. Within this framing, 'the human' in its ideal form was white and western.[12] Now shaping the thinking and activities of Silicon Valley, Gebru and Torres critique their inability to recognize and address the immediate harms and dangers that AI is posing to marginalized and underserved communities and people around the world.

More immediately, the call from Silicon Valley's leading minds for all AI research to pause is also a clear statement that no one but us should be at the forefront of the development of this powerful, humanity-transforming technology. Or, put differently, we are the unquestioned custodians of the future of all human life.

For global governance, however, the emphasis should be on the unique role that multilateralism and international regulation can play in ensuring the just and equitable distribution of the benefits of AI, rather than on long-term issues that only an elite few have the privilege to worry about. Global governance should take the lead in working to lower the cost of access to AI in countries of the majority world and in exploring how a taxation system for Big Tech and the new class of tech billionaires could support a major redistribution of wealth. Indeed, the World Inequality Report of 2022 makes it clear that 'addressing the challenges of the 21st century is not feasible without significant redistribution of income and wealth inequalities'.[13]

For redistribution to happen, there needs to be proper representation of a wide range of southern voices on global decision-making panels. Southern actors have less bargaining power in these spaces, yet at the same time the spaces themselves are vital for bringing balance to the global distribution of power and for ensuring that the needs and perspectives of the less powerful states are heard. But the representation of southern voices needs to be supported by evidence. Part of the reason why these people have less sway in global debates is that they are drowned out by the cacophony of western experience and of evidence and documents that support the western actors in these spaces. Together, this body of evidence can appear like a majority and consensus position, silencing the voices from the South. If the majority world can more forcefully claim its power in these spaces, then the West will, after all, be outnumbered. Understood in this way, the work of organizations like the African Observatory on Responsible AI, which attempts to produce evidence for, and support, African leaders in global decision-making, becomes crucial.

A Different Kind of Leader

I have been privileged enough to meet many astounding people from all corners of the world who are developing AI tools to help uplift their communities and societies. They are committed to using this technology for good rather than just for profit. And what is 'good' is defined by what they see is needed and by where they see AI has a distinct role to play.

In Nigeria, Nneka Mobisson is building AI technologies to address inequities in the healthcare system. In this she is driven by the passing of her father from a disease that could have been prevented. In Pakistan, Mariam Mustafa is leading the development of a voice-assisted app for overburdened clinicians to generate and keep medical records. In Uganda, a group of women at the Institute for Infectious Diseases at the University of Makerere are developing AI drones that offer antiretrovirals (ARVs) to people living with HIV/AIDs on the remote islands of Lake Victoria. In India, Amrita Mahale is developing an AI chatbot for teenage girls to ask questions about sexual and reproductive health, away from the stigma of local clinics. In South Africa, Leonora Tima is leading a group of women in building an AI-driven app to help women who are victims of domestic violence to navigate the complexities of the legal and justice system. In Tanzania, Joyce Nakatumba-Nabende is creating an application for local farmers to receive advice on crop diseases, yields, and local markets to sell their produce. In a country where maize is the staple food and represents half the production in agriculture, this technology offers the potential to contribute to food security more widely.

These people, and the teams they work with and lead, are striving against all odds to make phenomenal breakthroughs in designing and developing technologies that are fit for local purposes and tailored to local needs. They listen to what the communities they try to develop technologies for have to say in response to the question 'Is this working for you?' These are the kinds of innovations and approaches that can make lives better and easier for the people who most need it; and many of them are led by women.

Yet it is men who, by and large, reign over the empire of AI. Not simply 'men', as in a gender binary, but a ruthless masculinity that seeks to build at scale and dominate, reducing the social and human dimensions of data to units of calculability and prediction. The Cameroonian anthropologist Divine Fuh calls this the original sin of AI. Sam Altman's bid for trillions of dollars, the creation of new mega industries designed to serve the production of AI, or Elon Musk's trite promises of a world of abundance that AI will bring about – all these exemplify the towering self-conceit that smacks of male ego. In a recent book, Mustapha Suleyman lists money, national pride, and ego as the incentives that drive the AI industry.[14] This lethal combination has delivered us the empire of AI.

We will need a different kind of leadership to dismantle this empire. This leadership will need to understand, and be able to respond to, the social complexities of AI and the uncertainties our collective future holds. It will need to cherish and champion AI applications that are not scaled but are local and effective, like those described above.

In the history of development and peacekeeping, research has demonstrated the value of feminist leadership for communities and societies that overcome major social and political challenges. Some of the early work in my career was on transitional justice in postwar countries in Africa – on understanding how war-torn places recovered, revived themselves, and built just futures out of the ruins of violence and distrust. In particular, women like Ellen Johnson Sirleaf, who became the president of Liberia and Africa's first female president, were critical cornerstones in these efforts. Feminist perspectives and leadership were attuned to recognizing invisible work, thinking of and planning for unintended impacts on people and groups at the margins, and also for future generations. All of this contributed a deeper and more sustainable peace.[15]

These findings are not just confined to development and peacekeeping. Irene van Staveren has examined gender differences in the finance and banking sectors. Her findings are that gender stereotypes significantly constrain women's ability to reach leadership positions in these fields. But when they do reach these positions, having broken the glass ceiling, they often perform better than their male counterparts. In explanation, Staveren indicates that these differences 'seem to relate to the distinction between contextual ethics, concerned with relationships, flexibility . . . and self-discipline, that was found to be more related to women than to men in the empirical literature'.[16]

If we are serious about an equitable, just, and sustainable future with AI, we must recognize and promote feminist approaches in AI leadership and champion diversity in the development, rollout, and governance of AI technologies everywhere. Not only will this allow us to enact much broader foresight, as AI is designed and used in the world, but it is the only way we can work towards ensuring that these technologies benefit everyone everywhere.

Coda

The New Politics of Revolution

Much of what I have described in this book points to a world where divisions are deepening at an accelerating rate. They are, and we cannot ignore this reality. Nor can we give up on our commitments to build a sustainable, just, and equitable world where everyone everywhere can live a prosperous and dignified life.

At the 2024 World Economic Forum a narrative about 'the end of development' persisted, emanating from the Forum's own 2024 Global Risks Report.[1] The narrative holds that, with the rise of advanced technology and the related economic patterns that are set to continue – such as expeditious growth in technically endowed countries and industries, job losses and onshoring, restructured global supply chains, all coupled with the increasing climate burden in the global South – the trajectory of a deepening divide between rich and poor countries is beyond repair. 'Catching up' is no longer possible; it is not an agenda worth pursuing.[2]

But the fatigue and capitulation that may be surrounding the rise of global inequality cannot become entrenched and shape how we react. There are two key points that are forgotten in this complacent reaction. The first is that we are all implicated, and soon enough everyone everywhere will be affected. The second is that none of this is inevitable: there is always room for human collective action to change and divert a trajectory into which we may appear to be locked. We will explore these points in turn.

This Will Affect Us All

Throughout the world, where inequality rises, democracy weakens. If people are unable to meet their basic needs, they will migrate to places where they believe their needs can be met, compounding the already unwelcoming migration systems. With this, populism will rise, as people compete for limited jobs and state resources in a world managed by AI. And, where resources are scarce and inequality is rampant, criminality and even terrorism fester.

These conditions are heightened by the unsustainable business practices of the burgeoning AI industry, which contribute significantly to environmental degradation. Climate-related disasters, including ones that cause water scarcity and arid lands, will lead thousands to leave their homes and many more to bear the brunt of a regime from which they see no benefits. According to the World Bank, water scarcity alone is responsible for 10 per cent of the rise in global migration.[3]

But even now we are all becoming ensnared by AI, slaves to the machine. The more we use AI and the more we allow it to dominate our societies, the more we need it to fix – at scale – the problems it has, at scale, contributed to creating. We are dependent on AI technologies to detect when a photo or a video has been digitally altered or generated by another AI, while large-scale AI technologies are needed to track the AI-driven spread of misinformation online. What is more, we sense that something is not happening. We keep asking for research on the long-term impacts of our dependency on generative AI technologies like ChatGPT, aware that something of what it means to be human is being lost as their ubiquity grows. But we are wilfully ignoring the fact that, as we wait for this research to take place, we are all guinea pigs in the experiment.

According to the Chinese science fiction writer Chen Qiufan, we are racing towards an inevitable implosion, blinded by a shiny new technology and its promises of a brave new world. In conversations, he expressed a belief that some significant event of collapse or destruction was on the horizon, and that this might be the only way for us to realize that something needs to change.

Significantly, inequality breeds tension. The more inequality rises, the more internal tensions within our shared world space will boil up, and eventually erupt. This cannot, and should not, continue.

The Time for Action is Now

Understanding the historical antecedents of AI's power is crucial to resisting it, holding it to account, and building strategies towards a new and fundamentally equal world order. But old ways of thinking will not alone solve the oppressive effects of the novel forms and expressions of the power of AI today.

This is a basic aspect of decolonial thought and practice. First we identify the forms and structures of oppressive power at work in the world today – the forms and structures that deny inclusive and dignified futures to people marginalized through colonial histories. Second, we dismantle them. And, third, we build a new world order, in which everyone can truly flourish.

Calls to decolonize AI have come from all over the world, championed by critical scholars of AI or by indigenous groups; all, from Hawaii to New Zealand, practise and develop AI techniques on their own terms. These calls echo a growing global awareness that we need to attend to the debris left by the European empires of modernity, and also to respond to increasing instances of injustice and discrimination that arise in the AI industry and its discipline and practice and bear the mark of racial inequality and settler dispossession. But they demand action, and not just thought.

As a political strategy, decoloniality offers a further horizon. It opens up the possibility of transcending – collectively – colonialism and the mighty form of power it institutionalized. This new world into which the decolonial enterprise seeks to launch us must look quite different from the world we inhabit now. Indeed, it must involve a new form of human government, for – as we will explore here – the machine of liberal democracy has run its course; complicit in slavery and colonialism, liberal democracy cannot provide the profound human equality and freedom that decoloniality demands. In this way decoloniality is a radical invitation to the human imagination to conceive of a new world and build it.

Central to our response is a reclamation of human agency and creativity. It is from the human imagination and spirit, from the compassion of communities working together and from actions aligned to values of global humanity that we can begin to make amends and, collectively, change the trajectory before us. For Achille Mbembe, one of the world's foremost postcolonial thinkers, if there is to be a world renaissance, it will arise from Africa, where new life has been born from spaces of unlivability.[4] Indeed, we must look to new places for new solutions.

For the rest of us, we all have a part to play in resisting the dominance of AI and Big Tech. We can treat with scepticism the narratives they espouse. We can choose not to buy into their services where alternatives exist. We can support groups that seek to hold these companies accountable for the divisions they are sowing and the lives they are harming. We can demand that our governments put better regulation in place to address the imbalance of power. We can insist that our governments don't renege on their commitments to building a just and equitable world. We can all do our part. And we all should, because sooner or later we will all be impacted.

Notes

Note to Prologue

1 Bethan McKernan and Harry Davies, '"The Machine Did It Coldly": Israel Used AI to Identify 37000 Hamas Targets'. *The Guardian*, 3 April 2024. https://www.theguardian.com/world /2024/apr/03/israel-gaza-ai-database-hamas-airstrikes.

Notes to Introduction

1 PricewaterhouseCoopers (PwC), 'Sizing the Prize: What's the Real Value of AI for Your Business and How Can You Capitalise?'. PwC. 2017. https://www.pwc.com/gx/en/issues/analytics/assets /pwc-ai-analysis-sizing-the-prize-report.pdf.

2 Nishant Yonzan, Christoph Lakner, and Daniel Gerszon Mahler, 'Projecting Global Extreme Poverty up to 2030: How Close Are We to World Bank's 3% Goal?'. World Bank Blogs, 9 October 2020.

3 Lucas Chancel, Thomas Piketty, Emmanuel Saez, and Gabriel Zucman, *World Inequality Report 2022* (World Inequality Lab, 2022).

4 E. Tendayi Achiume, 'Racial Discrimination and Emerging Digital Technologies: A Human Rights Analysis: Report of the

Special Rapporteur on Contemporary Forms of Racism, Racial Discrimination, Xenophobia and Related Intolerance to the Forty-Fourth Session of the United Nations Human Rights Council'. 18 June 2020. A/HRC/44/57.

5 Ruha Benjamin, *Race after Technology* (Polity, 2019).

6 Safiya Umoja Noble, *Algorithms of Oppression: How Search Engines Reinforce Racism* (NYU Press, 2018).

7 Simone Browne, *Dark Matters: On the Surveillance of Blackness* (Duke University Press, 2015).

8 SPR, 'Women in Tech Statistics: 73% Experience Gender Bias in the Workplace'. 6 February 2024.

9 One of the most powerful pieces on this argument is by Argentinian postcolonial theorist Maria Lugones: Maria Lugones, 'The Coloniality of Gender', in W. Harcourt (ed.), *The Palgrave Handbook of Gender and Development* (Palgrave Macmillan, 2016).

10 Scholars have raised the question of colonialism in relation to the will of Big Tech to commodify all human experience as data points that feed insatiable AI systems. Nick Couldry and Ulises Mejias, for example, speak of this as 'data colonialism', exploring how human experience is being appropriated as capital in an increasingly data-driven market: Nick Couldry and Ulises Ali Mejias, *The Costs of Connection: How Data Is Colonizing Human Life and Appropriating It for Capitalism* (Stanford University Press, 2019).

11 Willie James Jennings, *The Christian Imagination: Theology and the Origins of Race* (Yale University Press, 2010).

12 I borrow this term from the Peruvian postcolonial thinker Aníbal Quijano.

13 As the theoretical foundations of the notion of a global South developed, important emphasis came to be placed on how the conditions and experiences of the global South existed within marginalized pockets of the global North – broadly, among immigrants and people of colour, who carry a connection with the long histories of slavery and colonialism.

14 Shahidul Alam, 'Majority World: Challenging the West's Rhetoric of Democracy'. *Amerasia Journal* 34.1 (2008): 87–98.
15 Thomas Piketty, *Capital in the Twenty-First Century* (Harvard University Press, 2014).

Notes to Chapter 1

1 Republic of South Africa, Department of Telecommunications and Postal Services, 'Notice 764 of 2018'. *Government Gazette*, 4 December 2018. https://www.gov.za/sites/default/files/gcis_document/201812/42078gen764.pdf.
2 J. McCarthy, M. L. Minsky, N. Rochester, and C. E. Shannon Bell, 'A Proposal for the Dartmouth Summer Research Project on Artificial Intelligence'. 31 August 1955. http://jmc.stanford.edu/articles/dartmouth/dartmouth.pdf.
3 Nick Bostrom, *Superintelligence: Paths, Dangers, Strategies* (Oxford University Press, 2014).
4 'Pause Giant AI Experiments: An Open Letter'. Future of Life Institute, 22 March 2023. https://futureoflife.org/open-letter/pause-giant-ai-experiments.
5 Clemency Burton-Hill, 'The Superhero of Artificial Intelligence: Can This Genius Keep It in Check?'. *The Guardian*, 16 February 2016. https://www.theguardian.com/technology/2016/feb/16/demis-hassabis-artificial-intelligence-deepmind-alphago.
6 Visit https://datagovhub.elliott.gwu.edu/germany-ai-strategy.
7 House of Lords, *AI in the UK: Ready, Willing and Able?* HL Paper 100, 2018. https://publications.parliament.uk/pa/ld201719/ldselect/ldai/100/100.pdf.
8 Visit https://digichina.stanford.edu/work/full-translation-chinas-new-generation-artificial-intelligence-development-plan-2017.
9 Stanford University Human-Centered Artificial Intelligence, *AI Index Report*. HAI 2023. https://hai.stanford.edu/research/ai-index-report.
10 Visit https://trumpwhitehouse.archives.gov/presidential-actions/executive-order-maintaining-american-leadership-artificial-intelligence.

11 Amber Kak and Sarah Myers West, 'AI Now 2023 Landscape: Confronting Tech Power'. AI Now: Institute, 11 April 2023. https://www.ainowinstitute.org/2023-landscape.

12 Bostrom, *Superintelligence*, p. 102.

13 Alex W. Palmer, '"An Act of War": Inside America's Silicon Blockade against China'. *New York Times*, 12 July 2023. https://www.nytimes.com/2023/07/12/magazine/semiconductor-chips-us-china.html?auth=login-google1tap&login=google1tap.

14 OECD.AI, 'Executive Order on Maintaining American Leadership in AI'. 2019. https://oecd.ai/en/dashboards/policy-initiatives/http:%2F%2Faipo.oecd.org%2F2021-data-policy Initiatives-24277.

15 Christy DeSmith, 'Why China Has Edge on AI, What Ancient Emperors Tell Us about Xi Jinping'. *Harvard Gazette*, 16 March 2023. https://news.harvard.edu/gazette/story/2023/03/why-china-has-an-edge-on-artificial-intelligence.

16 Valentin Hofmann, Pratyusha Ria Kalluri, Dan Jurafsky, and Sharese King. 'Dialect Prejudice Predicts AI Decisions about People's Character, Employability, and Criminality'. Cornell University, 1 March 2024. https://arxiv.org/abs/2403.00742.

17 Khari Johnson, 'How Wrongful Arrests Based on AI Derailed 3 Men's Lives'. *Wired*, 7 March 2022. https://www.wired.com/story/wrongful-arrests-ai-derailed-3-mens-lives.

18 David Theo Goldberg, *The Racial State* (Blackwell, 2002).

19 Crain Soudien, *A History of Race* (forthcoming).

20 Sabelo Ndlovu-Gatsheni, 'Perhaps Decoloniality Is the Answer? Critical Reflections on Development from a Decolonial Epistemic Perspective'. *Africanus* 43.2 (2013). https://hdl.handle.net/10520/EJC142701.

21 Tshepo Madlingozi, 'The Proposed Amendment to the South African Constitution: Finishing the Unfinished Business of Decolonisation?'. *Critical Legal Thinking*, 6 April 2018. https://criticallegalthinking.com/2018/04/06/the-proposed-amendment-to-the-south-african-constitution.

22 Benjamin, *Race after Technology*, p. 44.

23 Moustafa Bayoumi, 'The US Government Is Deploying Robot Dogs to the Mexico Border: Seriously?'. *The Guardian*, 14 February 2022. https://www.theguardian.com/commentisfree/2022/feb/14/us-government-deploying-robot-dogs-to-mexico-border.

24 GCHQ, 'Pioneering a New National Security: The Ethics of Artificial Intelligence'. https://www.gchq.gov.uk/files/GCHQAI Paper.pdf.

25 Filip Noubel, 'Why Do Western Governments Delegate Border Control to AI More and More? An Interview with Petra Molnar'. *Global Voices*, 5 April 2023. https://globalvoices.org/2023/04/05/why-do-western-governments-delegate-border-control-to-ai-more-and-more-an-interview-with-petra-molnar.

26 Petra Molnar, *The Walls Have Eyes* (New Press, 2024).

27 Ryan Gallagher and Ludovica Jona, 'We Tested Europe's New Lie Detector for Travelers – and Immediately Triggered a False Positive'. *The Intercept*, 26 July 2019. https://theintercept.com/2019/07/26/europe-border-control-ai-lie-detector.

28 Constanza M. Vidal Bustamante et al., 'Should Machines Be Allowed to "Read Our Minds"? Uses and Regulation of Biometric Techniques That Attempt to Infer Mental States'. *MIT Science Policy Review*, 29 August 2022. doi: 10.38105/spr.qy2iibrk72.

Notes to Chapter 2

 1 McKinsey, 'The Economic Potential of Generative AI'. Report, 14 June 2023. https://www.mckinsey.com/capabilities/mckinsey-digital/our-insights/the-economic-potential-of-generative-ai-the-next-productivity-frontier.

 2 UN Sustainable Development Goals, 'No Poverty: Why It Matters'. 2018. https://www.un.org/sustainabledevelopment/wp-content/uploads/2018/09/Goal-1.pdf.

 3 Oxfam International, 'Survival of the Richest'. 16 January 2023. https://www.oxfam.org/en/research/survival-richest and https://www.oxfam.org/en/press-releases/richest-1-bag-nearly-twice-much-wealth-rest-world-put-together-over-past-two-years.

4 J. Clement, 'US Digital Economy Gross Output 2005–2022'. *Statista*, 20 February 2024.

5 Wondwosen Tamrat, 'Bridging Digital Gender Gap through Inclusive STEM Education'. *University World News*, 25 May 2023. https://www.universityworldnews.com/post.php?story=20230523213059437#:~:text=State%20of%20gender%20digital%20divide&text=Women%20in%20this%20region%20are,the%20largest%20gender%20gap%20globally.

6 Visit https://openai.com/charter.

7 'Imperial Ambitions'. *Economist*, 9 April 2016.

8 Stanford University Human-Centered Artificial Intelligence, *AI Index Report*.

9 We will explore the problematics associated with insufficient data in Chapter 5.

10 Visit https://dbpedia.org/page/Cloud_Constellation.

11 Rachel Coldicutt and Tom McGrath, 'How Will AI Large Language Models Shape the Future and What Is the Right Regulatory Approach?'. Written Evidence to the House of Lords Communications and Digital Select Committee Inquiry: Large Language Models, 4 September 2023. https://committees.parliament.uk/writtenevidence/124251/html.

12 Hee Eun Lee, Woo Young Kim, Hyo Kang, and Kangwook Han, 'Still Lacking Reliable Electricity from the Grid, Many Africans Turn to Other Sources'. Afrobarometer Dispatch No. 514, 8 April 2022.

13 Dadabhai Naoroji, *Poverty of India* (Ranima Union Press, 1876).

14 Piketty, *Capital in the Twenty-First Century*, p. 121.

15 Sizwe Mpofu-Walsh, *The New Apartheid: Apartheid Did Not Die, It Was Privatised* (Tafelberg, 2021).

16 Fanny Pigeaud and Ndongo Samba Sylla, *Africa's Last Colonial Currency: The CFA Franc Story* (Pluto, 2021). The subsequent quotations are both from p. 3.

17 Walter Rodney, *How Europe Underdeveloped Africa* (Bogle-L'Ouverture, 1972).

18 Ibid., p. 208.

19 The AI global supply chain is discussed in detail in the next chapter.

20 This question of how fit for purpose AI is outside the West and Far East is the focus of Chapter 5.

Notes to Chapter 3

1 John Parker, *Making the Town: Ga State and Society in Early Colonial Accra* (ABC-CLIO, 2000).

2 World Economic Forum, 'A New Circular Vision for Electronics: Time for a Global Reboot'. 24 January 2019. https://www3.we forum.org/docs/WEF_A_New_Circular_Vision_for_Electronics .pdf.

3 Ibid., p. 14.

4 I. Issah, J. Arko-Mensah, T. Agyekum, D. Dwomoh, and J. Fobil, 'Health Risks Associated with Informal Electronic Waste Recycling in Africa: A Systematic Review'. *International Journal Environmental Research and Public Health* 19.21 (2022). doi: 10.3390/ijerph192114278.

5 Laura Vyda, 'Highest Level of World's Most Toxic Chemicals Found in African Free-Range Eggs: European e-Waste Dumping a Contributor'. *Arnika*, 24 April 2019. https://arnika.org/en/ news/most-toxic-chemicals-in-african-eggs.

6 World Economic Forum, 'A New Circular Vision for Electronics'.

7 Kate Crawford and Vladan Joler, 'Anatomy of an AI System: The Amazon Echo as an Anatomical Map of Human Labor, Data and Planetary Resources'. AI Now Institute and Share Lab, 7 September 2018. https://anatomyof.ai. See also Kate Crawford, *Atlas of AI* (Yale University Press, 2021).

8 Kwame Nkrumah, *Neo-Colonialism, the Last Stage of Imperialism* (Thomas Nelson & Sons, 1966). The book was published on the eve of a coup d'état in Ghana which resulted in the summary removal of Nkrumah from office, and his replacement by Joseph Ankrah, leader of the National Liberation Council.

9 Ibid., p. 200.

10 Ibid., p. 199.

Notes to Chapter 4

1 Visit https://sdgs.un.org/goals/goal8.

2 Olivia Solon, 'More than 70 per cent of US Fears Robots Taking over Our Lives, Survey Finds'. *The Guardian*, 4 October 2017. https://www.theguardian.com/technology/2017/oct/04/robots-artificial-intelligence-machines-us-survey.

3 B. Roberts, S. Gordon, J. Struwig, N. Bohler-Muller, and M. Gastrow, 'Promise or Precarity? South African Attitudes towards the Automation Revolution'. *Development Southern Africa* 39.11 (2022): 1–18.

4 Goldman Sachs, 'The Potentially Large Effects of Artificial Intelligence on Economic Growth'. 26 March 2023. https://www.gspublishing.com/content/research/en/reports/2023/03/27/d64e052b-0f6e-45d7-967b-d7be35fabd16.html.

5 As discussed in Chapter 1.

6 Rodney, *How Europe Underdeveloped Africa*, p. 287.

7 United Nations Industrial Development Organization, *Industrial Development Report 2020*. 2020. https://www.unido.org/sites/default/files/files/2019-12/UNIDO%20IDR20%20main%20report.pdf.

8 Jen Breman, 'A Short History of the Informal Economy'. *Global Labour Journal* 14.1 (2023), p. 23.

9 Daniel Motaung, 'How a Man from the Free State Came to Take on Facebook'. Research ICT Africa, 1 December 2023. https://researchictafrica.net/2023/12/01/daniel-motaung-how-a-man-from-the-free-state-came-to-take-on-facebook.

10 Josh Dzieza, 'AI Is a Lot of Work: As the Technology Becomes Ubiquitous, a Vast Tasker Underclass Is Emerging – and Not Going Anywhere'. *New York* magazine and *The Verge*, 20 June 2023. https://nymag.com/intelligencer/article/ai-artificial-intelligence-humans-technology-business-factory.html.

11 Gregory Clark and Robert C. Feenstra, 'Technology in the Great Divergence', in Michael D. Bordo, Alan M. Taylor, and Jeffrey G. Williamson (eds.), *Globalization in Historical Perspective* (University of Chicago Press, 2003).

Notes to Chapter 5

1 Human Rights Watch, *Automated Neglect: How the World Bank's Push to Allocate Cash Assistance Using Algorithms Threatens Rights* (Human Rights Watch, 2023).

2 Jean Drèze and Amartya Sen, *An Uncertain Glory: India and Its Contradictions* (Princeton University Press, 2013).

3 Visit https://x.com/theintercept/status/1669125864460566529.

4 Visit https://indiastack.org.

5 Such as those we explored in the final section of Chapter 1.

6 Access Now, *Bodily Harms: Mapping the Risks of Emerging Biometric Tech* (Access Now, 2023). https://www.accessnow.org/wp-content/uploads/2023/10/Bodily-harms-mapping-the-risks-of-emerging-biometric-tech.pdf.

7 Ibid.

8 Eileen Guo and Adi Renaldi, 'Deception, Exploited Workers, and Cash Handouts: How Worldcoin Recruited Its First Half a Million Test Users'. *MIT Technology Review*, 6 April 2022. https://www.technologyreview.com/2022/04/06/1048981/worldcoin-cryptocurrency-biometrics-web3.

9 For further information on the case, see Joseph Tinarwo and Suresh Chandra Babu, 'Chinese Artificial Intelligence in Africa: Digital Colonisation or Liberalisation?'. Tayarisha Working Paper 12, 2023.

10 Keith Breckenridge, *Biometric State* (Cambridge University Press, 2014), p. 12.

11 Arjun Appadurai, 'Number in the Colonial Imagination', in Carol A. Breckenridge and Peter van der Veer (eds.), *Orientalism and the Postcolonial Predicament: Perspectives on South Asia* (University of Pennsylvania Press, 1993), p. 334.

12 Browne, *Dark Matters*.

13 Chancel et al., World Inequality Report 2022.

14 Jean Drèze and Amartya Sen, *An Uncertain Glory*.

15 Ibid.

Notes to Chapter 6

1 Thomas Friedman, 'Our New Promethean Moment'. *New York Times*, 12 April 2023.

2 Mustafa Suleyman, *The Coming Wave* (Penguin, 2023).

3 See the table on levels of AI from Google DeepMind included in Chapter 4.

4 On the discussion of this letter led by the Future of Life Institute, see also Chapter 1.

5 Visit https://futureoflife.org/open-letter/pause-giant-ai-experi ments.

6 See Chloe Xiang, 'The Open Letter to Stop "Dangerous" AI Race Is a Huge Mess'. *Vice*, 29 March 2023. https://www.vice.com/en/article/qjvppm/the-open-letter-to-stop-dangerous-ai-race-is-a-huge-mess.

7 Visit https://commoncrawl.github.io/cc-crawl-statistics/plots /languages.html.

8 Emily M. Bender, Timnit Gebru, Angelina McMillan-Major, and Shmargaret Shmitchell, 'On the Dangers of Stochastic Parrots: Can Language Models Be Too Big?', in *Proceedings of the 2021 ACM Conference on Fairness, Accountability, and Transparency* (Association for Computing Machinery, 2021), pp. 610–622, here p. 613. https://s10251.pcdn.co/pdf/2021-bender-parrots.pdf.

9 Ibid., p. 614.

10 Gregory Gondwe, 'ChatGPT and the Global South: How Are Journalists in Sub-Saharan Africa Engaging with Generative AI?'. *Online Media and Global Communication* 2.2 (2023): 228–249. https://doi.org/10.1515/omgc-2023-0023.

11 Ibid., p. 241.

12 Ibid., p. 242. King Leopold II was the notorious Belgian colonial ruler of the Democratic Republic of Congo mentioned in Chapter 3 in relation to the country's colonial exploits into mining.

13 Ibid.

14 *The Continent* 64, 13 November 2021. https://www.thecon tinent.org/_files/ugd/287178_6c92ada5e8be4af9841db421a3ed4 be8.pdf.

15 An important work here is Edward Said, *Culture and Imperialism* (Vintage, 1993).

16 Ngũgĩ wa Thiong'o, *Decolonising the Mind* (Boydell & Brewer, 1986), p. 108.

17 Rockie Sibanda, 'Mother-Tongue Education in a Multilingual Township: Possibilities for Recognising lok'shin Lingua in South Africa'. *Reading & Writing* 10.1 (2019). https://files.eric.ed.gov /fulltext/EJ1228689.pdf.

18 R. Chandler et al., 'The Impact of COVID-19 among Black Women: Evaluating Persp,ectives and Sources of Information'. *Ethnicity and Health* 26.1 (2021): 80–93.

19 Caroline Criado Perez, *Invisible Women: Exposing Data Bias in a World Designed for Men* (Chatto & Windus, 2019).

20 Kehinde Oluwaseun Omotoso, Jimi Adesina, and Ololade G. Adewole, 'Exploring Gender Digital Divide and Its Effect on Women's Labor Market Outcomes in South Africa'. *African Journal of Gender, Society and Development* 9.4 (2020). https://hdl.handle.net/10520/ejc-aa_jgida1-v9-n4-a5.

Notes to Chapter 7

1 Anna Jobin, Marcello Ienca, and Effy Vayena, 'The Global Landscape of AI Ethics Guidelines'. *Nature Machine Intelligence* 1 (2019): 388–399.

2 Shakir Mohamed, Marie-Therese Png, and William Isaac, 'Decolonial AI: Decolonial Theory as Sociotechnical Foresight in Artificial Intelligence'. *Philosophy and Technology* 33 (2020): 659–684, here p. 675.

3 Sabelo Mhlambi, 'From Rationality to Relationality: Ubuntu as an Ethical and Human Rights Framework for Artificial Intelligence Governance'. Carr Centre Discussion Paper, 8 July 2020.

4 See *The Continent* 142 (2023); for this particular story, visit https://www.reddit.com/user/TheContinentAfrica/comments /17xpv65/the_continent_issue_142.

5 Gayatri Spivak, *A Critique of Postcolonial Reason* (Harvard University Press, 1999).

6 Achille Mbembe, *Critique of Black Reason* (Duke University Press, 2017).

7 Olayinka Akanle, Gbenga S. Adejare, and Jojolola Fasuyi, 'To What Extent Are We All Humans? Of Culture, Politics, Law and LBGT Rights in Nigeria', in Melissa Styen and William Mpofu (eds.), *Decolonising the Human* (Wits University Press, 2021), p. 49.

8 See Helen Tilley, *Africa as a Living Laboratory: Empire, Development, and the Problem of Scientific Knowledge, 1870–1950* (University of Chicago Press, 2011).

9 Christopher Wylie, *Mindf*ck: Inside Cambridge Analytica's Plot to Break the World* (Profile Books, 2019), p. 69.

10 Joe Parkinson, Nicholas Bariyo, and Josh Chin, 'Huawei Technicians Helped African Governments Spy on Political Opponents'. *Wall Street Journal*, 15 August 2019.

11 Laís Martins, 'Unleash the bololô: Masses of Delivery Workers Set Off Horns and Fireworks at Bad Customers' Homes'. 4 December 2003. https://restofworld.org/2023/bololo-delivery-workers-fireworks-horns-protest.

12 See Angela Saini, *Superior: The Return of Race Science* (Fourth Estate, 2019).

13 Chancel et al., *World Inequality Report 2022*.

14 Mustafa Suleyman, *The Coming Wave* (Penguin, 2023).

15 See e.g. Wendy Harcourt (ed.), *Feminist Perspectives on Sustainable Development* (Zed Books, 1994).

16 Irene van Staveren, 'The Lehman Sisters Hypothesis'. Working Paper No 545, ISS, June 2012, here p. 19. (Reprinted in *Cambridge Journal of Economics* 38 (2014): 995–1014.)

Notes to Coda

1 World Economic Forum, *Global Risks Report 2024*. World Economic Forum, 10 January 2024.

2 Teniola T. Tayo, 'The African Union vs "the End of Development"'. Africa Policy Research Institute, 28 February 2024. https://afripoli.org/the-african-union-vs-the-end-of-development.

3 World Bank, 'Going with the Flow: Water's Role in Global Migration'. 23 August 2021.

4 Achille Mbembe, *Brutalism* (Duke University Press, 2024).

Key Readings

Access Now. *Bodily Harms: Mapping the Risks of Emerging Biometric Tech* (Access Now, 2023).

E. Tendayi Achiume. 'Racial Discrimination and Emerging Digital Technologies: A Human Rights Analysis'. Report of the Special Rapporteur on Contemporary Forms of Racism, Racial Discrimination, Xenophobia and Related Intolerance to the Forty-Fourth Session of the United Nations Human Rights Council. 18 June 2020. A/HRC/44/57.

Arjun Appadurai. 'Number in the Colonial Imagination', in Carol A. Breckenridge and Peter van der Veer (eds.), *Orientalism and the Postcolonial Predicament: Perspectives on South Asia* (University of Pennsylvania Press, 1993).

Chinmayi Arun. 'AI and the Global South: Designing for Other Worlds', in Markus D. Dubber, Frank Pasquale, and Sunit Das (eds.), *Oxford Handbook of Ethics of AI* (Oxford University Press, 2020).

Moustafa Bayoumi. 'The US Government Is Deploying Robot Dogs to the Mexico Border: Seriously?'. *The Guardian*, 14 February 2022. https://www.theguardian.com/commentisfree/2022/feb/14/us-government-deploying-robot-dogs-to-mexico-border.

Emily M. Bender. Timnit Gebru, Angelina McMillan-Major, and Shmargaret Shmitchell. 'On the Dangers of Stochastic Parrots:

Can Language Models Be Too Big?', in *Proceedings of the 2021 ACM Conference on Fairness, Accountability, and Transparency* (Association for Computing Machinery, 2021).

Ruha Benjamin. *Race after Technology: Abolitionist Tools for the New Jim Code* (Polity, 2019).

Abeba Birhane. 'Algorthmic Colonization of Africa'. *SCRIPTed* 17.2 (2020): 389–409.

Nick Bostrom. *Superintelligence: Paths, Dangers, Strategies* (Oxford University Press, 2014).

Keith Breckenridge. *Biometric State: The Global Politics of Identification and Surveillance in South Africa, 1850 to the Present* (Cambridge University Press, 2014).

Jen Breman. 'A Short History of the Informal Economy'. *Global Labour Journal* 14.1 (2023): 21–39.

Simone Browne. *Dark Matters: On the Surveillance of Blackness* (Duke University Press, 2015).

Joy Buolamwini and Timnit Gebru. 'Gender Shades: Intersectional Accuracy Disparities in Commercial Gender Classification', in *Proceedings of Machine Learning Research* (Association for Computing Machinery, 2018).

Aimé Césaire. *Discourse on Colonialism*, trans. Joan Pinkham (Monthly Review Press, 2001).

Lucas Chancel, Thomas Piketty, Emmanuel Saez, and Gabriel Zucman. *World Inequality Report 2022* (World Inequality Lab, 2022).

Gregory Clark and Robert C. Feenstra. 'Technology in the Great Divergence', in Michael D. Bordo, Alan M. Taylor, and Jeffrey G. Williamson (eds.), *Globalization in Historical Perspective* (University of Chicago Press, 2003).

Nick Couldry and Ulises Ali Mejias. *The Costs of Connection: How Data Is Colonizing Human Life and Appropriating It for Capitalism* (Stanford University Press, 2019).

Caroline Criado Perez. *Invisible Women: Exposing Data Bias in a World Designed for Men* (Vintage, 2019).

Kate Crawford. *Atlas of AI: Power, Politics, and the Planetary Costs of Artificial Intelligence* (Yale University Press, 2022).

Kate Crawford and Vladan Joler. 'Anatomy of an AI System: The Amazon Echo as an Anatomical Map of Human Labor, Data and Planetary Resources'. AI Now Institute and Share Lab, 7 September 2018. https://anatomyof.ai.

Jean Drèze and Amartya Sen. *An Uncertain Glory: India and Its Contradictions* (Princeton University Press, 2013).

Josh Dzieza. 'AI Is a Lot of Work: As the Technology Becomes Ubiquitous, a Vast Tasker Underclass Is Emerging – and Not Going Anywhere'. *New York Times* and *The Verge*, 20 June 2023. https://nymag.com/intelligencer/article/ai-artificial-intelligence -humans-technology-business-factory.html.

Virginia Eubanks. *Automating Inequality: How High Tech Tools Profile, Police, and Punish the Poor* (Picador, 2018).

Michel Foucault. *The Will to Knowledge*, vol. 1 of his *History of Sexuality* (Penguin, 1978).

Ryan Gallagher and Ludovica Jona. 'We Tested Europe's New Lie Detector for Travellers – and Immediately Triggered a False Positive'. *The Intercept*, 26 July 2019. https://theintercept.com/ 2019/07/26/europe-border-control-ai-lie-detector.

David Theo Goldberg. *The Racial State* (Blackwell, 2002).

Gregory Gondwe. 'ChatGPT and the Global South: How Are Journalists in Sub-Saharan Africa Engaging with Generative AI?'. *Online Media and Global Communication* 2.2 (2023). https://doi .org/10.1515/omgc-2023-0023.

Ramón Grosfoguel. 'The Epistemic Decolonial Turn'. *Cultural Studies* 21.2–3 (2007): 211–223.

Eileen Guo and Adi Renaldi. 'Deception, Exploited Workers, and Cash Handouts: How Worldcoin Recruited Its First Half a Million Test Users'. *MIT Technology Review*, 6 April 2022. https://www .technologyreview.com/2022/04/06/1048981/worldcoin-crypto currency-biometrics-web3.

Wendy Harcourt (ed.). *Feminist Perspectives on Sustainable Development* (Zed Books, 1994).

Valentin Hofmann, Pratyusha Ria Kalluri, Dan Jurafsky, and Sharese King. 'Dialect Prejudice Predicts AI Decisions about People's

Character, Employability, and Criminality'. Cornell University, 1 March 2024. https://arxiv.org/abs/2403.00742.

Human Rights Watch. *Automated Neglect: How the World Bank's Push to Allocate Cash Assistance Using Algorithms Threatens Rights* (Human Rights Watch, 2023).

Khari Johnson. 'How Wrongful Arrests Based on AI Derailed 3 Men's Lives'. *Wired*, 7 March 2022. https://www.wired.com/story/wrongful-arrests-ai-derailed-3-mens-lives.

Amber Kak and Sarah Myers West. 'AI Now: 2023 Landscape: Confronting Tech Power'. AI Now Institute, 11 April 2023. https://www.ainowinstitute.org/2023-landscape.

U. Kalpagam. 'The Colonial State and Statistical Knowledge'. *History of the Human Sciences* 13.2 (2000). https://doi.org/10.1177/09526950022120665.

U. Kalpagam. *Rule by Numbers: Governmentality in Colonial India* (Rowman & Littlefield, 2014).

Os Keyes. 'The Misgendering Machines: Trans/HCI Implications of Automatic Gender Recognition', in *Proceedings of the ACM on Human–Computer Interaction*, vol. 2 (2018). https://doi.org/10.1145/3274357.

Vili Lehdonvirta. *Cloud Empires: How Digital Platforms Are Overtaking the State and How We Can Regain Control* (MIT Press, 2022).

Tshepo Madlingozi. 'The Proposed Amendment to the South African Constitution: Finishing the Unfinished Business of Decolonisation?'. *Critical Legal Thinking*, 6 April 2018. https://criticallegalthinking.com/2018/04/06/the-proposed-amendment-to-the-south-african-constitution.

Vukosi Marivate et al. 'Investigating an Approach for Low Resource Language Dataset Creation, Curation and Classification: Setswana and Sepedi', in *Proceedings of the First Workshop on Resources for African Indigenous Languages*. ELRA, 2020.

Achille Mbembe. *Critique of Black Reason* (Duke University Press, 2017).

Achille Mbembe. *Brutalism* (Duke University Press, 2024).

Sabelo Mhlambi. 'From Rationality to Relationality: Ubuntu as an Ethical and Human Rights Framework for Artificial Intelligence Governance'. Carr Centre Discussion Paper, 8 July 2020.

Shakir Mohamed, Marie-Therese Png, and William Isaac. 'Decolonial AI: Decolonial Theory as Sociotechnical Foresight in Artificial Intelligence'. *Philosophy and Technology* 33 (2020): 659–684.

Daniel Motaung. 'How a Man from the Free State Came to Take on Facebook'. Research ICT Africa, 1 December 2023. https://rese archictafrica.net/2023/12/01/daniel-motaung-how-a-man-from-the-free-state-came-to-take-on-facebook.

Walter Mignolo and Catherine Walsh. *On Decoloniality* (Duke University Press, 2018).

Petra Molnar. *The Walls Have Eyes* (New Press, 2024).

Sizwe Mpofu-Walsh. *The New Apartheid: Apartheid Did Not Die, It Was Privatised* (Tafelberg, 2021).

Dadabhai Naorojim. *Poverty of India* (Ranima Union Press, 1876).

Sabelo Ndlovu-Gatsheni. 'Perhaps Decoloniality Is the Answer? Critical Reflections on Development from a Decolonial Epistemic Perspective'. *Africanus* 43.2 (2013). https://hdl.handle.net/10520 /EJC142701.

Kwame Nkrumah. *Neo-Colonialism, the Last Stage of Imperialism* (Thomas Nelson & Sons, 1966).

Safiya Umoja Noble. *Algorithms of Oppression: How Search Engines Reinforce Racism* (New York University Press, 2018).

Fanny Pigeaud and Ndongo Samba Sylla. *Africa's Last Colonial Currency: The CFA Franc Story* (Pluto, 2021).

Thomas Piketty. *Capital in the Twenty-First Century* (Harvard University Press, 2014).

Hannah Ritchie. *Not the End of the World: How We Can Be the First Generation to Build a Sustainable Planet* (Chatto & Windus, 2024).

Walter Rodney. *How Europe Underdeveloped Africa* (Bogle-L'Ouverture, 1972).

Edward Said. *Orientalism* (Pantheon, 1978).

Edward Said. *Culture and Imperialism* (Vintage, 1993).

Angela Saini. *Superior: The Return of Race Science* (Fourth Estate, 2019).

Klaus Schwab. *The Fourth Industrial Revolution* (World Economic Forum, 2016).

Crain Soudien. *Realising the Dream: Unlearning the Logic of Race in the South African School* (Human Sciences Research Council, South Africa, 2012).

Mustafa Suleyman. *The Coming Wave* (Penguin, 2023).

Teniola T. Tayo. 'The African Union vs "the End of Development"'. Africa Policy Research Institute, 28 February 2024. https:// afripoli.org/the-african-union-vs-the-end-of-development.

Ngũgĩ wa Thiong'o. *Decolonising the Mind* (Boydell & Brewer, 1986).

Helen Tilley. *Africa as a Living Laboratory: Empire, Development, and the Problem of Scientific Knowledge, 1870–1950* (University of Chicago Press, 2011).

Joseph Tinarwo and Suresh Chandra Babu. 'Chinese Artificial Intelligence in Africa: Digital Colonisation or Liberalisation?'. Tayarisha Working Paper 12, 2023.

Eve Tuck and K. Wayne Yang. 'Decolonization Is Not a Metaphor'. *Decolonization: Indigeneity, Education & Society* 1.1 (2012): 1–40.

David Wallace-Wells. *The Uninhabitable Earth: A Story of the Future* (Penguin, 2020).

World Economic Forum. *Global Risks Report 2024*. World Economic Forum, 10 January 2024.

Chloe Xiang. 'The Open Letter to Stop "Dangerous" AI Race Is a Huge Mess'. *Vice*, 29 March 2023. https://www.vice.com/en/article/ qjvppm/the-open-letter-to-stop-dangerous-ai-race-is-a-huge-mess.

Shoshana Zuboff. *The Age of Surveillance Capitalism: The Fight for a Human Future at the New Frontier of Power* (Profile Books, 2018).

Index

African Observatory on Responsible
AI 175
Agbogbloshie 80–2, 84, 112
Alexa 84, 102, 105
AlphaGo 30–1, 35
Altman, Sam 133, 155–6, 176
Apple 16, 26, 64, 88, 101, 173
Amazon 16, 26, 62, 84–8, 101, 106,
109–10, 115
apartheid 19–20, 49, 71, 93, 109,
110, 135
artificial general intelligence
(AGI) 31–3, 36, 59–60,
105–6
artificial intelligence (AI)
arms race 26–7, 36–8, 41–6,
62–3, 156–7
bias in 4, 8, 10, 12–4, 26, 46–50,
134, 143–7, 152, 161–2
definition of 2–3, 27–8
environmental impact of 66–8,
94–9
history of 27–34
investment in 29, 37, 39
neural networks 28, 30, 33, 60–1,
117, 144, 159

research and development (R&D)
in 28–9, 38
Atlantic slave trade 16, 49, 57, 74, 90

Belgian colonialism 91–4, 147
Big Tech 10, 16, 25–6, 54, 55, 58–9,
65–8, 99, 119, 125, 138, 133, 166,
168–9, 171–2, 174, 181
BioIDs 131–6
Bostrom, Nick 31–3, 41
Brazil 85–9, 127, 152, 170
British Empire 16–8, 70–1, 79, 90–1,
109, 164–5

Cambridge Analytica 24, 64, 165–6
Cameroon 68, 75, 157, 176
Cape Town 101, 105, 109–10, 121
cash transfers 127–30
ChatGPT 28, 33–4, 36, 47, 54,
58–61, 67, 84, 97, 104, 116,
138–42, 146–47, 151, 157, 172,
180
Chile 83, 95–7, 169
China 5, 15, 25, 26, 34, 36–7, 38–46,
56, 62–3, 74, 85, 89, 97, 135,
146, 167–8

CloudWalk 134–5, 167
Covid-19 5, 65, 111, 114, 124, 151

Dartmouth College 27–8
data centres 61–2, 66–7, 69, 84, 97
data labelling 117–20
decolonization 20, 71–2, 80, 91–3, 149, 162, 180–1
Defense Advanced Research Projects Agency (DARPA) 28–9, 73
Democratic Republic of Congo 83, 85, 88–94, 96, 146–7, 169
democracy, impact of AI on 21, 45, 161, 165, 179–80

effective altruism 173–4
Elon Musk 18, 36, 124, 141, 173, 176
Ethiopia 85, 148, 161
eugenics 49, 173–4
European colonialism 14–18, 21, 40, 46, 48–9, 57–8, 69–75, 83, 89–94, 104, 108–12, 124, 136–7, 164–5
e-waste 80–2
European Union 35, 36, 53, 82

Facebook 10, 64–5, 118, 148, 165
facial recognition technologies 13–14, 47, 52, 131, 134–5, 157, 159
Foucault, Michel 146–7
Fourth Industrial Revolution 23–5
France 25, 45, 57, 72
    French colonialism 72, 91, 123
Future of Life Institute 33, 41, 173

Gebru, Timnit 142–3, 173–4
Germany 25, 35, 45, 51, 101, 142
Ghana 77, 79–82, 92, 112
gig work(er) 8, 113–15, 118–22, 169
globalization 15, 19, 104, 108, 112, 135

Google 13, 16, 30, 35, 57, 62, 64, 77, 101, 143, 150–1
Google DeepMind 18, 26, 30–1, 34–5, 75, 105–6, 140–1

Hassabis, Demis 34, 124, 141
Huawei 39, 168
human rights 5, 8, 12, 53, 88, 99, 127, 166–9
Human Rights Watch 127–30

India 70, 85, 104, 107, 121, 131–2, 135, 137–8, 152, 175
    Aadhaar 131–2
indigenous groups 16, 46, 73, 88, 95–6, 98, 109–11, 135, 137, 180
Internet, the 23, 29–30, 56–7, 60, 62, 68–9, 76, 77, 84, 117, 120, 133, 142–3, 150, 152–3
    governance of 172
    shutdowns 168

Japan 25, 34, 45, 101, 108
job loss(es) 4, 54, 103–5, 108–9, 157, 178
Jordan 127–30, 161

Kenya 10, 56–7, 77, 117–19, 124, 126, 133–4, 149
killer robots 103, 159

large language models (LLMs) 32–3, 47, 54, 61, 63, 67, 103, 116, 119, 139–45, 148–50, 152–3, 157, 170
Lumumba, Patrice 92

M-Pesa 56–7, 77, 124
McCarthy, John 27
Meta 36, 65, 117–18, 140, 142, 170
Microsoft 16, 26, 60–2, 64, 75, 88, 116, 139, 171
migrants
    refugees and asylum seekers
    migrant workers

Motuang, Daniel 10, 118–19
Myanmar 64, 148

Nigeria 57, 85, 126, 166, 175
Nkrumah, Kwame 80, 91–3

OpenAI 33, 59–61, 67, 116, 133, 139,
　　142, 144, 155

platform economies 63–5, 113–15,
　　119, 169–70

race and racism 12–14, 27, 46–53,
　　181
racial bias in AI 135–6, 143–5,
　　161–2, 173
rare earth elements (REEs) 85,
　　88–97
Rwanda 85, 89, 124

semiconductor chips 42–3, 60–1,
　　155
Silicon Valley 60, 116, 126, 174
Siri 77, 102, 106, 144
South Africa 8–10, 19–20, 24–5, 40,
　　48–9, 57, 68, 71, 90–1, 93–4,
　　103, 104–5, 110–11, 112–13, 121,
　　135–6, 164–5, 175–6
South Korea 31, 34, 35
supercomputer 36, 61–3

superintelligence 31–3, 36, 41, 105–6
surveillance 13, 43, 49–50, 52, 114,
　　131–4, 136, 159–60
Sustainable Development Goals
　　(United Nations) 18, 20–1, 40,
　　102, 172

Takaful 127–30, 161
TikTok 41–2, 44
Turing, Alan 27, 35, 51

undersea cables 67–8, 61
UNESCO 123, 171–2
United Kingdom 25, 35–7, 43–4, 51,
　　61–3, 100, 130, 137, 152, 160,
　　165
United States 12, 25, 34–7, 41–6, 54,
　　56, 61, 65, 80, 102–4, 126, 148,
　　152, 160, 165
Uyghur people 45, 167–8

women 107, 112–13
　　and digital divide around gender
　　57, 152–3
　　impact of AI and technology on
　　50, 105, 124, 142
　　in AI 13–14, 175–7
Worldcoin 133–4

Zimbabwe 91, 101, 134–5, 162, 167